Construction Analytics

Mohsen Shahandashti
Bahram Abediniangerabi • Ehsan Zahed
Sooin Kim

Construction Analytics

Forecasting and Investment Valuation

Mohsen Shahandashti
The Department of Civil Engineering
The University of Texas at Arlington
Arlington, TX, USA

Bahram Abediniangerabi
The Department of Civil Engineering
The University of Texas at Arlington
Arlington, TX, USA

Ehsan Zahed
The Department of Civil Engineering
The University of Texas at Arlington
Arlington, TX, USA

Sooin Kim
The Department of Civil Engineering
The University of Texas at Arlington
Arlington, TX, USA

ISBN 978-3-031-27294-3   ISBN 978-3-031-27292-9   (eBook)
https://doi.org/10.1007/978-3-031-27292-9

© The Editor(s) (if applicable) and The Author(s), under exclusive license to Springer Nature Switzerland AG 2023

This work is subject to copyright. All rights are solely and exclusively licensed by the Publisher, whether the whole or part of the material is concerned, specifically the rights of translation, reprinting, reuse of illustrations, recitation, broadcasting, reproduction on microfilms or in any other physical way, and transmission or information storage and retrieval, electronic adaptation, computer software, or by similar or dissimilar methodology now known or hereafter developed.

The use of general descriptive names, registered names, trademarks, service marks, etc. in this publication does not imply, even in the absence of a specific statement, that such names are exempt from the relevant protective laws and regulations and therefore free for general use.

The publisher, the authors, and the editors are safe to assume that the advice and information in this book are believed to be true and accurate at the date of publication. Neither the publisher nor the authors or the editors give a warranty, expressed or implied, with respect to the material contained herein or for any errors or omissions that may have been made. The publisher remains neutral with regard to jurisdictional claims in published maps and institutional affiliations.

This Springer imprint is published by the registered company Springer Nature Switzerland AG
The registered company address is: Gewerbestrasse 11, 6330 Cham, Switzerland

Contents

1 Introduction to Construction Analytics 1
   1.1 The Impacts and Challenges of the Construction Industry 1
   1.2 Construction Analytics for Solving Construction Industry
       Challenges 2
   1.3 Construction Analytics Techniques: Forecasting
       and Investment Valuation 3
   1.4 R Language for Implementing Construction Analytics 4
   1.5 Summary 5
   1.6 Exercise Problems 5
   References 5

**2 Construction Time Series Forecasting Using Univariate Time
Series Models** 7
   2.1 Introduction 7
   2.2 Construction Cost Time Series 8
       2.2.1 Example 1: Highway Construction Spending (HCS) 8
       2.2.2 Example 2: National Highway Construction Cost
             Index (NHCCI) 8
       2.2.3 Example 3: Composite Index of Iowa Highway
             Construction (IHC) 10
       2.2.4 Example 4: California Construction Cost Index (CCCI) ... 10
   2.3 Characteristics of Time Series 11
       2.3.1 Stationarity 11
       2.3.2 Seasonality 14
       2.3.3 Trend 16
   2.4 Univariate Time Series Forecasting 16
       2.4.1 Moving Average (MA) 18
       2.4.2 Autoregressive (AR) 18
       2.4.3 Exponential Smoothing (ES) 22
       2.4.4 Autoregressive Moving Average (ARMA) 25

		2.4.5	Autoregressive Integrated Moving Average (ARIMA)	27
		2.4.6	Seasonal Autoregressive Integrated Moving Average (SARIMA)...	29
	2.5	Diagnostic Tests for Time Series Models		35
		2.5.1	Diagnostic Tests for No Autocorrelation	35
		2.5.2	Diagnostic Tests for Homoscedasticity	37
		2.5.3	Diagnostic Tests for Normality	39
	2.6	Summary		41
	2.7	Problems		41
	References...			42
3	**Construction Forecasting Using Time Series Volatility Models**			45
	3.1	Introduction		45
	3.2	Time Series Volatility Models		46
		3.2.1	Autoregressive Conditional Heteroscedasticity (ARCH)...	46
		3.2.2	Generalized Autoregressive Conditional Heteroscedasticity (GARCH)	51
	3.3	Diagnostic Tests for Time Series Volatility Models............		53
		3.3.1	Results of Diagnostic Tests for No Autocorrelation	53
		3.3.2	Results of Diagnostic Tests for Homoscedasticity........	54
	3.4	Estimating Volatility Using ARCH and GARCH Models		55
	3.5	Forecasting the TFC Time Series Using ARIMA-ARCH and ARIMA-GARCH Models...............................		56
	3.6	Summary		59
	3.7	Exercise Problems		60
	References...			62
4	**Construction Time Series Forecasting Using Multivariate Time Series Models**...			63
	4.1	Introduction		63
	4.2	Potential Explanatory Time Series.........................		64
		4.2.1	Stationarity......................................	65
		4.2.2	Granger Causality	66
		4.2.3	Cointegration Test	68
	4.3	Multivariate Forecasting Models		69
		4.3.1	Vector Autoregressive (VAR)........................	69
		4.3.2	Vector Error Correction (VEC)	69
		4.3.3	Forecasting Errors of VEC Models	71
	4.4	Summary		73
	4.5	Exercise Problems		73
	References...			74
5	**Construction Forecasting Using Recurrent Neural Networks**			75
	5.1	Introduction		75
	5.2	Recurrent Neural Networks		77

	5.3	Model Development Process	78
		5.3.1 Data Preparation	78
		5.3.2 Train and Test Datasets	79
		5.3.3 Data Rescaling	79
		5.3.4 RNN Parameters	80
	5.4	Simple RNNs	82
	5.5	Long Short-Term Memory (LSTM)	84
	5.6	Gated Recurrent Unit (GRU)	89
	5.7	Forecasting Errors of Time Series Models	91
	5.8	Summary	93
	5.9	Exercise Problems	93
	References		93
6	**Investment Valuation of Construction Projects Under Uncertainty**		95
	6.1	Stochastic Life-Cycle Cost Analysis of Construction Projects	95
	6.2	Life-Cycle Cost Comparison of Alternative Construction Projects	104
	6.3	Real Options Analysis of Construction Projects	110
	6.4	Binomial Tree Model	111
	6.5	Summary	122
	6.6	Exercise Problems	122
	References		124

Appendix A: Conventional Investment Valuation Techniques for Evaluating Construction Projects 125

Appendix B: R and R Package Installation and Helpful Communities 145

Appendix C 147

Appendix D 149

Appendix E 151

Appendix F 153

Appendix G 155

Appendix H 157

Appendix I 159

Appendix J 161

Appendix K 163

Appendix L 165

Appendix M 167

Appendix N 169

Appendix O 171

Index 185

Chapter 1
Introduction to Construction Analytics

Abstract The construction industry has a crucial impact on the economy. Data analytics provides a unique opportunity to improve construction decision-making, enhance construction productivity, and reduce construction cost overruns. Although data analytics have tremendous potential to improve strategic decision-making in the construction industry as an ever-increasing volume of data becomes available, it has not been fully exploited on a larger scale in the construction industry due to a lack of proper training and educational materials. Two powerful construction analytics techniques (i.e., forecasting and investment valuation) are introduced that can potentially help address several grand challenges in the construction industry. Advanced forecasting techniques can improve cost estimation accuracy and assist engineers in avoiding a bid loss or a profit loss. Investment valuation techniques assist engineers in identifying the appropriate time to invest, quantifying the investment risks in projects, and determining the optimum value of an investment for maximizing the returns on investments. This book provides theoretical explanations, hands-on practice problems with R code scripts, and exercises for learning the construction industry's most advanced and valuable data analytics techniques.

Keywords Construction analytics · Forecasting · Investment valuation · Strategic decision-making · R programming

1.1 The Impacts and Challenges of the Construction Industry

The construction industry plays a significant role in the global economy, with approximately $10 trillion in annual spending on construction-related goods and services (Barbosa et al., 2017). The cumulative value of global infrastructure investment is estimated to reach $94 trillion by 2040, including new construction and maintenance (an average of $3.8 trillion per year) (Oxford Economics, 2017).

Nearly one-third of global infrastructure investment is expected to occur in the United States (Barbosa et al., 2017). The construction industry is one of the major economic drivers in every state of the United States, creating nearly $1.4 trillion worth of structures every year, with more than 733,000 employers and more than 7 million employees (Lang, 2019; Simonson, 2022). The $1.2 trillion Bipartisan Infrastructure Bill, which has recently passed, is expected to increase employment by more than 250,000 construction jobs and further accelerate the overall economic growth by 3.2% annually (Zandi & Yaros, 2021). Despite the significant impact of the construction industry on the economy, this industry suffers from severe chronic problems, such as poor productivity, cost and time overruns, and rising material costs. Over the past two decades, the average productivity growth rate in the construction industry has only been 1 percent annually, while productivity has increased by 2.8 percent and 3.6 percent annually in the whole world economy and manufacturing industry, respectively (Barbosa et al., 2017). One of the main reasons for low productivity in the construction industry is the lack of or poor strategic decision-making (Turner et al., 2021).

1.2 Construction Analytics for Solving Construction Industry Challenges

Today, data analytics techniques support strategic decision-making in almost every industry, such as banking, healthcare, manufacturing, and oil and gas (Kudyba, 2019). Data analytics provides an unprecedented and unique opportunity to address grand challenges in the construction industry. For example, construction analytics can help improve cost estimation and investment valuation in a volatile construction market. For example, Kim et al. (2021) showed how construction cost estimators could implement data analytics to improve the accuracy of cost estimation and forecasting in pipeline construction projects to avoid bid failure, cost overruns, and profit loss. Construction engineers can diagnose and quantify post-disaster construction cost fluctuations and formulate risk mitigation plans (Khodahemmati & Shahandashti, 2020). Data analytics have tremendous potential to improve strategic decision-making in the construction industry as an ever-increasing volume of data becomes available (Bilal et al., 2016).

Despite the significant benefits of data analytics for strategic decision-making, the construction industry has not fully exploited the opportunities for leveraging advanced data analytics at scale (Akinosho et al., 2020; Barbosa et al., 2017; Turner et al., 2021). One of the significant challenges in implementing data analytics in the construction industry is the lack of proper training and educational materials. This is primarily a grand challenge for medium and small companies with limited resources for internal training and educational programs. Construction programs in higher education institutions are slowly rethinking their curriculum to include data analytics. However, there is no textbook with practical examples for teaching construction analytics to graduate and senior undergraduate construction students. This textbook addresses this enormous gap in construction education and practice. This book enables students to write R programming codes to implement advanced data analytics for solving practical construction problems.

1.3 Construction Analytics Techniques: Forecasting and Investment Valuation

Forecasting and investment valuation are powerful data analytics techniques to facilitate strategic decision-making for the success of a construction project. Accurate forecasting is essential for strategic decision-making (Fildes et al., 2019). The high complexity of the construction industry with a variety of stakeholders and influencing factors is one of the difficulties in accurately forecasting construction data (Gondia et al., 2020). The inaccurate construction cost forecasting can result in cost overruns, increasing the risks of project delay, project abandonment, financial losses, and even insolvency of the contractors (Signor et al., 2020). Advanced forecasting techniques enable engineers to forecast construction variables such as material costs and prepare risk management plans against future volatility. For example, a contractor who bids for a highway construction project can enhance cost estimation by forecasting the National Highway Construction Cost Index (NHCCI) along with other macroeconomic leading indicators such as crude oil price (Shahandashti & Ashuri, 2016). Advanced forecasting techniques can improve cost estimation accuracy and assist engineers in avoiding a bid loss or a profit loss. The forecasting techniques covered in this textbook include univariate time series forecasting, multivariate time series forecasting, volatility forecasting, and machine learning methods. These forecasting techniques enable engineers to analyze construction data and market conditions, forecast the construction variables including construction costs, develop more accurate budgets, and prepare more accurate winning bids.

Investment valuation is another prominent data analytics technique for strategic decision-making in construction projects (Grover et al., 2018). Investment valuation techniques assist engineers in identifying the appropriate time to invest, quantifying project investment risks, and determining the optimum value of an investment for maximizing the returns on investments. Conventional investment valuation methods are explained in Appendix A to prepare readers with no construction economics background to fully comprehend real options analysis for investment valuation of construction projects under uncertainty.

This book focuses on two valuable construction analytics techniques: forecasting and investment valuation. The book is organized into seven chapters. Chapters 2, 3, 4, and 5 discuss various data analytics techniques for forecasting construction variables. These chapters explain univariate time series forecasting, multivariate time series forecasting, volatility forecasting, and machine learning methods, accompanied by practical examples and R codes. Chapter 6 describes real options analysis for the investment valuation of construction projects under uncertainty. Each chapter explains the theoretical foundations of analytics techniques and illustrates them with practical construction examples. An R package and several R codes are provided to analyze the construction data, conduct statistical tests, and develop data-intensive models for solving practical construction examples. An R package titled "cdar" is created for implementing data analytics techniques for construction problems. The open-source R package "cdar" contains the necessary functions and datasets to work through numerous examples and solve the end-of-chapter exercise problems within R environment. The "cdar" package can be accessed and downloaded from GitHub. This package provides public datasets used in the examples. A detailed description of the package can be found in the package description using this link: https://github.com/Bahram-Abediniangerabi/cdar. R Code 1.1 can be used to install the "cdar" package on RStudio from GitHub using DevTools.

R Code 1.1

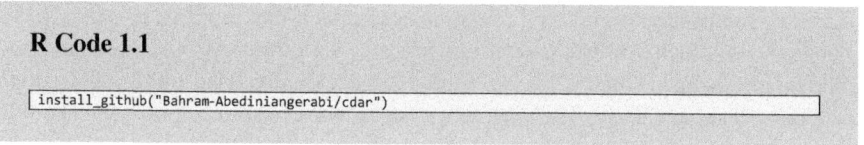

1.4 R Language for Implementing Construction Analytics

In this textbook, R language was selected for implementing construction analytics techniques for several reasons. First, R is an open-source programming language that can be freely downloaded and installed on your computer via the official website. R enables users to start and utilize a powerful programming language for data analytics at no cost. Second, R is already a commonly used programming language among researchers and practitioners for data analytics and statistical software development in academia and industry. Numerous valuable and reliable resources, such as tutorials, online tutors, R communities, and R forums, are freely available for users to get technical and customer support. Valuable resources, including the website for R installation, are provided in Appendix B. R codes can be executed in a user-friendly graphical interface (e.g., RStudio). Students with little to no programming background can quickly start using RStudio to run the codes and explore the results. Lastly, R is a powerful general-purpose programming language for numerous applications, including a wide variety of data analytics. Programming codes, statistical packages, and R functions have been actively developed by users worldwide in its open-source environment. R users can take advantage of this open-source environment and exploit numerous statistical modeling packages and algorithms for data analytics.

Construction analytics techniques can assist engineers in identifying potential risks, forecasting construction variables, determining the most profitable and economical strategy, and ultimately enhancing construction productivity (Pan & Zhang, 2021). This book provides theoretical explanations, hands-on practices, and exercise problems, along with R codes and datasets, for learning the most advanced and valuable data analytics techniques in the construction industry. This textbook enables readers to understand the necessary theoretical background of construction analytics and practice construction analytics techniques by running the R codes to solve the construction problems, such as cost overruns and poor investment timing.

This book also contains instructions to install and run required R packages and codes. The step-by-step R code scripts for actual practice and exercise problems encourage the readers to enjoy learning by doing and adopting the construction analytics techniques to address construction challenges. It is expected that this book will be a valuable reference for students, researchers, and data analysts in engineering, construction, and project management to improve their knowledge and skills for strategic decision-making with the recent advancements in construction analytics.

1.5 Summary

This chapter recommended two powerful construction analytics techniques (i.e., forecasting and investment valuation) for addressing some of the grand challenges in the construction industry. Advanced forecasting techniques enable engineers and practitioners to forecast construction variables, such as material costs, and enhance their cost estimation accuracy. Investment valuation techniques can assist engineers in determining the appropriate investment timing, project investment risks, and the optimum value of an investment for maximizing the returns on investments. This book provides theoretical explanations, hands-on practices with R code scripts, and exercise problems for learning the most advanced and valuable construction analytics techniques.

1.6 Exercise Problems

1. Review the report *Reinventing Construction: A Route of Higher Productivity* by Barbosa et al. (2017), and discuss the current problems and challenges that the construction industry faces.
2. Discuss how the forecasting of construction data can benefit a construction project and address the challenges of the construction industry.
3. Discuss how the investment valuation of a construction project can improve investments and strategic decision-making.
4. Download R and RStudio and install the "cdar" package on RStudio from GitHub using R Code 1.1.

References

Akinosho, T. D., Oyedele, L. O., Bilal, M., Ajayi, A. O., Delgado, M. D., Akinade, O. O., & Ahmed, A. A. (2020). Deep learning in the construction industry: A review of present status and future innovations. *Journal of Building Engineering, 32*, 101827. https://doi.org/10.1016/j.jobe.2020.101827

Barbosa, F., Woetzel, J., & Mischke, J. (2017). *Reinventing construction: A route of higher productivity*. McKinsey Global Institute.

Bilal, M., Oyedele, L. O., Qadir, J., Munir, K., Ajayi, S. O., Akinade, O. O., Owolabi, H. A., Alaka, H. A., & Pasha, M. (2016). Big Data in the construction industry: A review of present status, opportunities, and future trends. *Advanced Engineering Informatics, 30*(3), 500–521. https://doi.org/10.1016/j.aei.2016.07.001

Fildes, R., Ma, S., & Kolassa, S. (2019). Retail forecasting: Research and practice. *International Journal of Forecasting*, S016920701930192X. https://doi.org/10.1016/j.ijforecast.2019.06.004

Gondia, A., Siam, A., El-Dakhakhni, W., & Nassar, A. H. (2020). Machine learning algorithms for construction projects delay risk prediction. *Journal of Construction Engineering and Management, 146*(1), 04019085. https://doi.org/10.1061/(ASCE)CO.1943-7862.0001736

Grover, V., Chiang, R. H. L., Liang, T.-P., & Zhang, D. (2018). Creating strategic business value from Big Data analytics: A research framework. *Journal of Management Information Systems, 35*(2), 388–423. https://doi.org/10.1080/07421222.2018.1451951

Khodahemmati, N., & Shahandashti, M. (2020). Diagnosis and quantification of postdisaster construction material cost fluctuations. *Natural Hazards Review, 21*(3), 04020019. https://doi.org/10.1061/(ASCE)NH.1527-6996.0000381

Kim, S., Abediniangerabi, B., & Shahandashti, M. (2021). Pipeline construction cost forecasting using multivariate time series methods. *Journal of Pipeline Systems Engineering and Practice, 12*(3), 04021026. https://doi.org/10.1061/(ASCE)PS.1949-1204.0000553

Kudyba, S. (2019). *Big Data, mining, and analytics: Components of strategic decision making.* Taylor & Francis Group.

Lang, H. (2019, Dec 4). *Top industries in every state.* Stacker. https://stacker.com/stories/2571/top-industries-every-state.

Oxford Economics. (2017). *Global infrastructure outlook-infrastructure investment need 50 countries, 7 sectors to 2040.* Oxford Economics. https://cdn.gihub.org/outlook/live/methodology/Global+Infrastructure+Outlook+-+July+2017.pdf.

Pan, Y., & Zhang, L. (2021). Roles of artificial intelligence in construction engineering and management: A critical review and future trends. *Automation in Construction, 122*, 103517. https://doi.org/10.1016/j.autcon.2020.103517

Shahandashti, S. M., & Ashuri, B. (2016). Highway construction cost forecasting using Vector Error Correction Models. *Journal of Management in Engineering, 32*(2), 04015040. https://doi.org/10.1061/(ASCE)ME.1943-5479.0000404

Signor, R., Love, P. E. D., Marchiori, F. F., & Felisberto, A. D. (2020). Underpricing in social infrastructure projects: Combating the institutionalization of the *Winner's Curse*. *Journal of Construction Engineering and Management, 146*(12), 05020018. https://doi.org/10.1061/(ASCE)CO.1943-7862.0001926

Simonson, K. (2022). *Construction data.* The Associated General Contractors. https://www.agc.org/learn/construction-data.

Turner, C. J., Oyekan, J., Stergioulas, L., & Griffin, D. (2021). Utilizing Industry 4.0 on the construction site: Challenges and opportunities. *IEEE Transactions on Industrial Informatics, 17*(2), 746–756. https://doi.org/10.1109/TII.2020.3002197

Zandi, M., & Yaros, B. (2021). Macroeconomic consequences of the infrastructure investment and jobs act & build back better framework. *Moody's Analytics*.

Chapter 2
Construction Time Series Forecasting Using Univariate Time Series Models

Abstract Construction costs generally change over time. Therefore, forecasting construction costs is essential for effective cost management of construction projects. This chapter aims to introduce several construction time series variables, such as Highway Construction Spending and National Highway Construction Cost Index, and demonstrate procedures for investigating the characteristics of such time series. This chapter introduces several univariate time series forecasting models, such as moving average, autoregressive, and Holt-Winters exponential smoothing models. It also provides practical examples for training and using them for forecasting construction time series. The chapter also includes R Codes, solutions, model diagnostics, and interpretations of the solutions.

Keywords Construction costs · Univariate time series forecasting · Stationarity · Seasonality · Autocorrelation · Moving average · Autoregressive · Exponential smoothing · Autoregressive moving average · Autoregressive integrated moving average · Seasonal autoregressive moving average

2.1 Introduction

A series of data points indexed in time intervals is called time series (Brockwell & Davis, 2016). In other words, a time series consists of sequential points taken at successive time intervals. Time series Y is expressed by:

$$Y = \{Y_t : t \in T\} \tag{2.1}$$

where Y_t is the scalar measurement at time t, which belongs to the time index set, T. Time increments are commonly at regular time intervals, such as hour, day, week, month, and year. The sequence or order in the observations is a significant property of any time series that makes a time series different than another type of data. A time series is changed by shifting or modifying the order of observations. The construction industry generates a large volume of time-indexed data, such as construction cost time series.

This chapter introduces the principles of time series analytics and univariate time series forecasting, illustrated by developing forecasting models for several publicly available construction indexes. First, several construction cost time series are introduced. Second, fundamental statistical attributes of a time series, such as stationarity, seasonality, and trend, are explained and demonstrated on construction cost indexes. Third, univariate time series forecasting techniques, such as moving average and autoregressive methods, are described and used to forecast construction cost indexes. The required datasets can be accessed using the R package "Construction Analytics." Lastly, diagnostic tests for univariate time series forecasting techniques are discussed to examine if the modeling assumptions are satisfied.

2.2 Construction Cost Time Series

The Highway Construction Spending (HCS), National Highway Construction Cost Index (NHCCI), composite index of Iowa Highway Construction (IHC), and California Construction Cost Index (CCCI) are publicly available examples of construction cost time series. Construction cost indexes are mainly used for adjusting the cost estimation of capital infrastructure projects to avoid cost overruns (Kim et al., 2020). The following examples describe these time series:

2.2.1 Example 1: Highway Construction Spending (HCS)

The US Census Bureau publishes Highway Construction Spending (HCS) as the value of federal construction put in place (not seasonally adjusted). HCS is a monthly measure of the value of construction spending on highways and streets by the federal government (US Census Bureau, 2019). Figure 2.1 illustrates the monthly HCS time series for 16 years from 2003 to 2018. HCS raw data is available in Appendix C.

2.2.2 Example 2: National Highway Construction Cost Index (NHCCI)

The Federal Highway Administration (FHWA) publishes the quarterly price index of the National Highway Construction Cost Index (NHCCI), which measures the average changes in highway construction costs over time. The NHCCI is the conversion of the current-dollar highway construction expenditures to real-dollar expenditures (FHWA, 2019). Figure 2.1 illustrates the quarterly NHCCI time series for 16 years from 2003 to 2018. NHCCI raw data is available in Appendix D.

2.2 Construction Cost Time Series

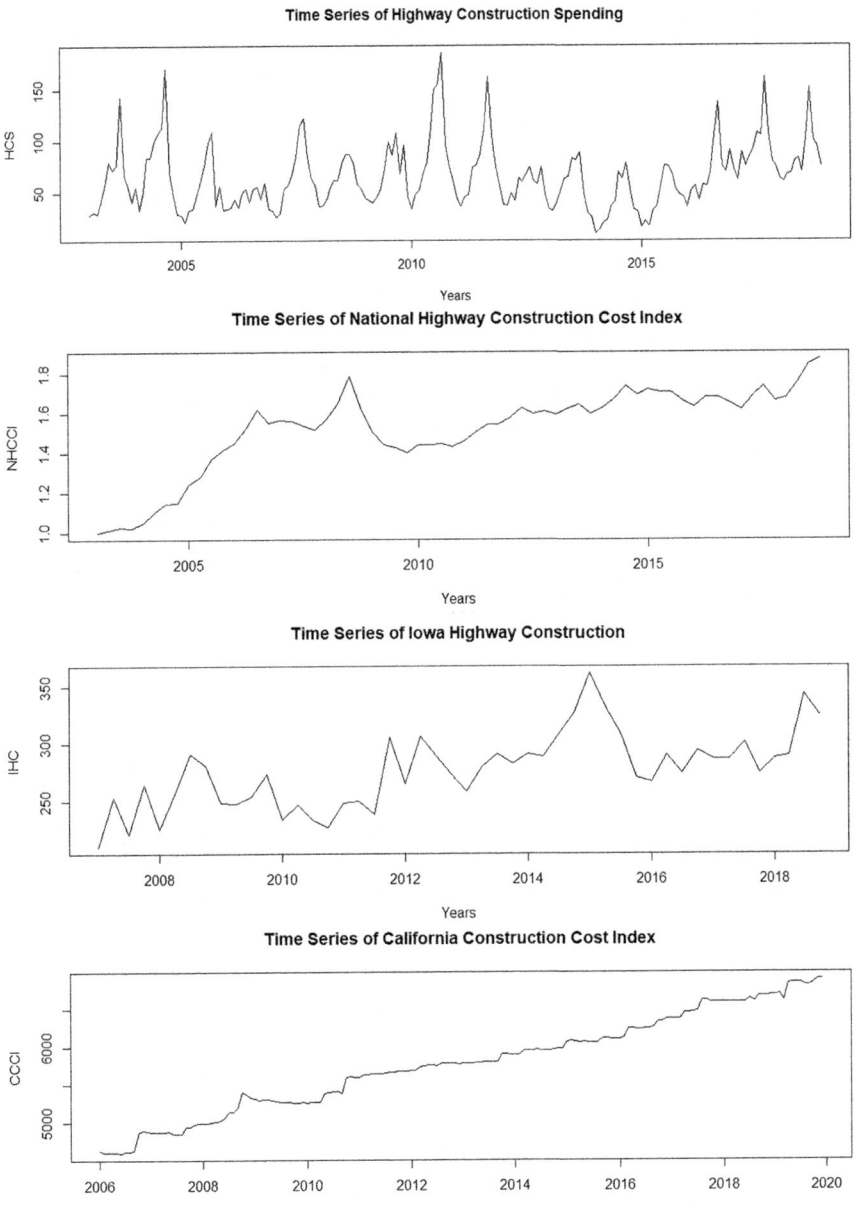

Fig. 2.1 Time series of HCS, NHCCI, IHC, and CCCI

2.2.3 Example 3: Composite Index of Iowa Highway Construction (IHC)

The composite index of Iowa Highway Construction (IHC) is a quarterly index published by the Iowa Department of Transportation (IOWADOT). The IHC index is the collection of the three construction categories of roadway on awarded contracts: excavation, surfacing, and structures. This index presents the weighted averages of the six indicator items (i.e., roadway, borrow, and embankment-in-place, HMA pavement and shoulder mixes, PCC pavements, reinforcing steel, structural steel, and structural concrete) on awarded contracts starting from 1986 (IOWADOT, 2019). Figure 2.1 illustrates the IHC time series for 12 years from 2007 to 2018. IHC raw data is available in Appendix E.

2.2.4 Example 4: California Construction Cost Index (CCCI)

California Construction Cost Index (CCCI) is the monthly index for San Francisco and Los Angeles developed locally based upon the average Building Cost Index (BCI) (California Department of General Services, 2019). Figure 2.1 illustrates the monthly CCCI time series for 14 years from January 2006 to December 2019. CCCI raw data is available in Appendix F.

R Code 2.1 can be used to plot the HCS, NHCCI, IHC, and CCCI time series.

R Code 2.1
Import and plot time series

```
library(cdar)
hcs <- cdar::hcs
hcs.ts <- ts(hcs, frequency = 12, start = c(2003,1), end =c(2018,12))
nhcci <- cdar::nhcci
nhcci.ts <- ts(nhcci, frequency = 4, start = c(2003,1))
ihc <- cdar::ihc
ihc.ts <- ts(ihc, frequency = 4, start = c(2007,1), end = c(2018,4))
ccci <- cdar::ccci
ccci.ts <- ts(ccci, frequency = 12, start = c(2006,1), end = c(2019,12))
plot.ts(hcs.ts, main = "Time Series of Highway Construction Spending", xlab = "Years", ylab = "HCS")
plot.ts(nhcci.ts, main = "Time Series of National Highway Construction Cost Index", xlab = "Years", ylab = "NHCCI")
plot.ts(ihc.ts, main = "Time Series of Iowa Highway Construction" , xlab = "Years", ylab = "IHC")
plot.ts(ccci.ts, main = "Time Series of California Construction Cost Index", xlab = "Years", ylab = "CCCI")
```

2.3 Characteristics of Time Series

Identifying the characteristics of a time series is critical to creating appropriate forecasting models. Stationarity, seasonality, and trend are the most important characteristics of a time series, which should be identified before developing any forecasting models.

2.3.1 Stationarity

The statistical properties of a stationary time series are constant over time, similar to those that have been time-shifted (Brockwell & Davis, 2016). Stationarity can be strict or weak. In a time series, if $(X_1, \ldots, X_n)$ and $(X_{1+h}, \ldots, X_{n+h})$ have the same joint distributions for all integers h and $n > 0$, the time series is strictly stationary (Heckert et al., 2002). It means that the mean, variance, and covariance are not the function of time for a strict stationary time series. A weak stationary time series, on the other hand, has the same mean at all time points, but there is a serial dependency, which means that it violates the statistical assumption of zero autocovariance (Palachy, 2019). Hereafter, the term stationary is used to indicate strictly stationary; if a process is stationary in the weak sense, we use the term weakly stationary.

The most common methods for examining the stationarity of a time series are the graphical method and statistical hypothesis testing. Autocorrelation is a measure of association between current and past data points in a time series. The autocorrelation function (ACF) is a visual representation method for determining the stationarity of a time series. Autocorrelation shows the similarity between a time series and its lagged version over consecutive time intervals. ACF plot checks weak stationarity.

The time series autocorrelation values tend to degrade to zero quickly in an ACF plot of a stationary time series, whereas the degradation happens more slowly for a nonstationary time series. Figure 2.2 shows the ACF test results for all the construction time series introduced in Sect. 2.2. The ACF results show a slow decaying pattern above the dotted blue lines (approximate confidence interval to judge the stationarity of the time series) for all the time series. Since all the time series cross the dotted blue line, it is concluded that they are not strictly stationary. However, HCS and IHC show weak stationarity.

The ACF plot for the time series is generated using R Code 2.2.

As shown in Fig. 2.2, the ACF test offers a useful graphical tool to decide whether a time series is stationary. However, these ACF plots do not determine the statistical significance of such decisions about a time series' stationarity. Statistical methods, such as unit root tests, can help determine whether a time series is stationary with a specific significance level. The augmented Dickey-Fuller (ADF) and Phillips-Perron tests are well-known unit root tests.

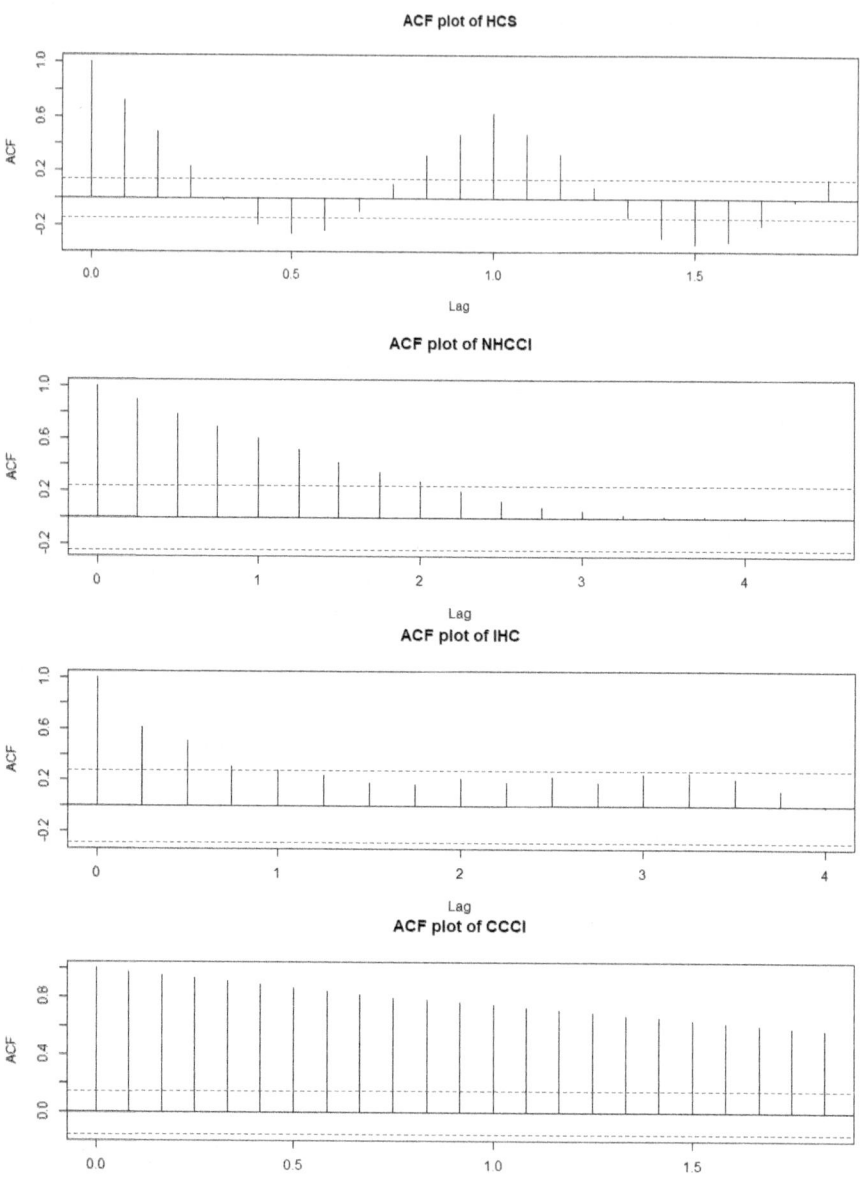

Fig. 2.2 ACF test results of HCS, NHCCI, IHC, and CCCI time series

2.3 Characteristics of Time Series

R Code 2.2
ACF test

```
acf(hcs.ts, type = c("correlation"), main = "ACF plot of HCS")
acf(nhcci.ts, type = c("correlation"), main = "ACF plot of NHCCI")
acf(ihc.ts, type = c("correlation"), main = "ACF plot of IHC")
acf(ccci.ts, type = c("correlation"), main = "ACF plot of CCCI")
```

The augmented Dickey-Fuller (ADF) test is one of the most popular statistical hypothesis testing methods for determining the stationarity of time series. The ADF test's null hypothesis is that the time series is nonstationary. Rejection of the ADF null hypothesis means that the time series is stationary. R Code 2.3 was used to test the stationarity of HCS, NHCCI, IHC, and CCCI time series using the ADF test. Schwert (2002) recommended selecting the lag order for the ADF test using the following equation:

$$k = \left\lfloor 12 \cdot \left(\frac{T}{100}\right)^{1/4} \right\rfloor \quad (2.2)$$

where T is the time series' length. Table 2.1 shows the results of the ADF test for HCS, NHCCI, IHC, and CCCI time series. The results show insufficient evidence to reject the ADF test's null hypothesis (non-stationarity) for all the time series with a 5% significance level. Therefore, HCS, NHCCI, IHC, and CCCI are nonstationary.

R Code 2.3
ADF test

```
library(tseries)
adf.test(hcs.ts, k=12*(length(hcs.ts)/100)^0.25)
adf.test(nhcci.ts, k=12*(length(nhcci.ts)/100)^0.25)
adf.test(ihc.ts, k=12*(length(ihc.ts)/100)^0.25)
adf.test(ccci.ts, k=12*(length(ccci.ts)/100)^0.25)
```

Stationarity is the first requirement for most statistical forecasting methods. Transformations, such as differencing, have been widely used to smooth the mean and variance of a time series by eliminating changes in the level of a time series and transform a nonstationary time series into a stationary one.

Table 2.1 ADF test results of HCS, NHCCI, IHC, and CCCI time series

Augmented Dickey-Fuller test
Data: hcs.ts Dickey-Fuller = −2.3634, lag order = 14.126, p-value = 0.4241 Alternative hypothesis: stationary
Data: nhcci.ts Dickey-Fuller = −3.1027, lag order = 10, p-value = 0.1282 Alternative hypothesis: stationary
Data: ihc.ts Dickey-Fuller = −2.1176, lag order = 10, p-value = 0.5272 Alternative hypothesis: stationary
Data: ccci.ts Dickey-Fuller = −2.6727, lag order = 14, p-value = 0.2955 Alternative hypothesis: stationary

The first difference of a time series is the series of changes from one period to the following period. For data points Y_t, the first difference is equal to $Y_t - Y_{t-1}$. The differencing operator is applied once on HCS, NHCCI, IHC, and CCCI time series to convert them to stationary time series using R Code 2.4. Figure 2.3 illustrates the first differenced HCS, NHCCI, IHC, and CCCI time series. ADF test is applied to the differenced time series to test whether the time series become stationary after applying the differencing operator. Table 2.2 shows the results of the ADF test for the first differenced time series. The results show that HCS and CCCI become stationary after applying the differencing operator once, while NHCCI and IHC time series remain nonstationary. The NHCCI and IHC time series may be differenced one more time to become stationary.

R Code 2.4
Applying difference operator on time series

```
hcs_diff = diff(hcs.ts)
nhcci_diff = diff(nhcci.ts)
ihc_diff = diff(ihc.ts)
ccci_diff = diff(ccci.ts)
```

2.3.2 Seasonality

Seasonality is quite typical in construction time series. Seasonality in a time series refers to the variations or fluctuations that repeat at specific regular intervals, such as weekly, monthly, or quarterly. The seasonal factor of a time series is the periodic component. Figure 2.4 illustrates the seasonal components of the HCS, NHCCI,

2.3 Characteristics of Time Series

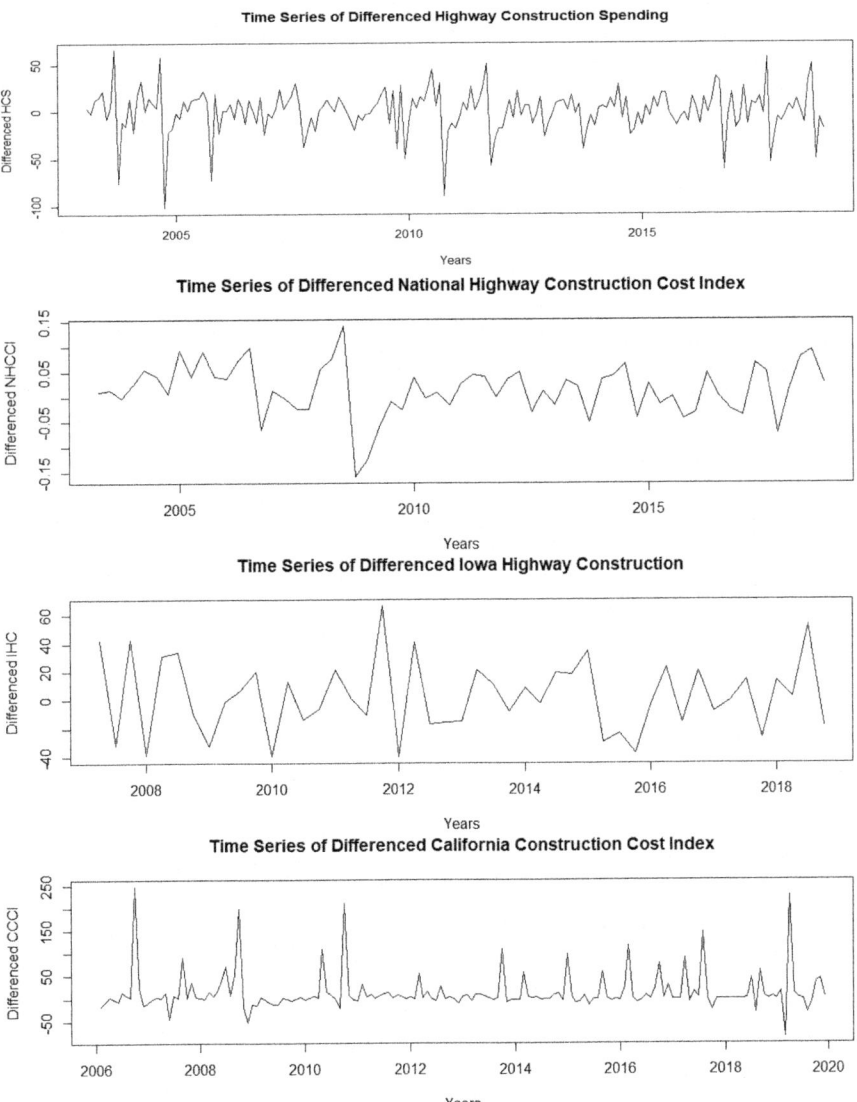

Fig. 2.3 Time series of the first differenced HCS, NHCCI, IHC, and CCCI time series

IHC, and CCCI time series. Figure 2.4 shows high seasonality for HCS, NHCCI, and CCCI but not for IHC. Therefore, a forecasting model capable of capturing seasonality is needed to deal with the seasonal component of these time series. For instance, Holt-Winters linear method could potentially be used for forecasting these time series since it can use the seasonality of the time series as the additive component. On the other hand, a simple exponential smoothing method may be adequate for forecasting IHC since it does not have seasonality.

Table 2.2 ADF results of first differenced HCS, NHCCI, IHC, and CCCI time series

Augmented Dickey-Fuller test
Data: hcs_diff Dickey-Fuller = −8.7094, lag order = 5, p-value = 0.01* Alternative hypothesis: stationary
Data: nhcci_diff Dickey-Fuller = −2.0426, lag order = 10.691, p-value = 0.5577 Alternative hypothesis: stationary
Data: ihc_diff Dickey-Fuller = −2.1367, lag order = 9.9359, p-value = 0.5196 Alternative hypothesis: stationary
Data: ccci_diff Dickey-Fuller = −3.8541, lag order = 13.641, p-value = 0.01805 Alternative hypothesis: stationary

*p-value smaller than printed p-value

R Code 2.5 can be used for decomposing and plotting time series seasonal components. Decomposing a time series into its components helps to understand a time series forecasting problem in terms of modeling complexity and how to best capture each of these components in a model.

2.3.3 Trend

The general, smooth, long-term, and average tendency of a time series is called a trend, which can be upward or downward, or a combination of both. This increase or decrease does not need to be in the same direction throughout a given period (Harvey, 1985). Trend components of HCS, NHCCI, IHC, and CCCI time series are shown in Fig. 2.5. The extracted trend component of HCS, NHCCI, IHC, and CCCI time series are plotted using R Code 2.6. Figure 2.5 shows that NHCCI and CCCI time series have upward trends, while the direction of trends for HCS and IHC time series changes over time.

2.4 Univariate Time Series Forecasting

Moving average, autoregressive, exponential smoothing, autoregressive moving average, and autoregressive integrated moving average models are univariate time series forecasting models used to forecast the construction time series. The following steps can be taken to model and forecast time series:

- Examining the time series' main characteristics, such as stationarity, trend, and seasonality

2.4 Univariate Time Series Forecasting

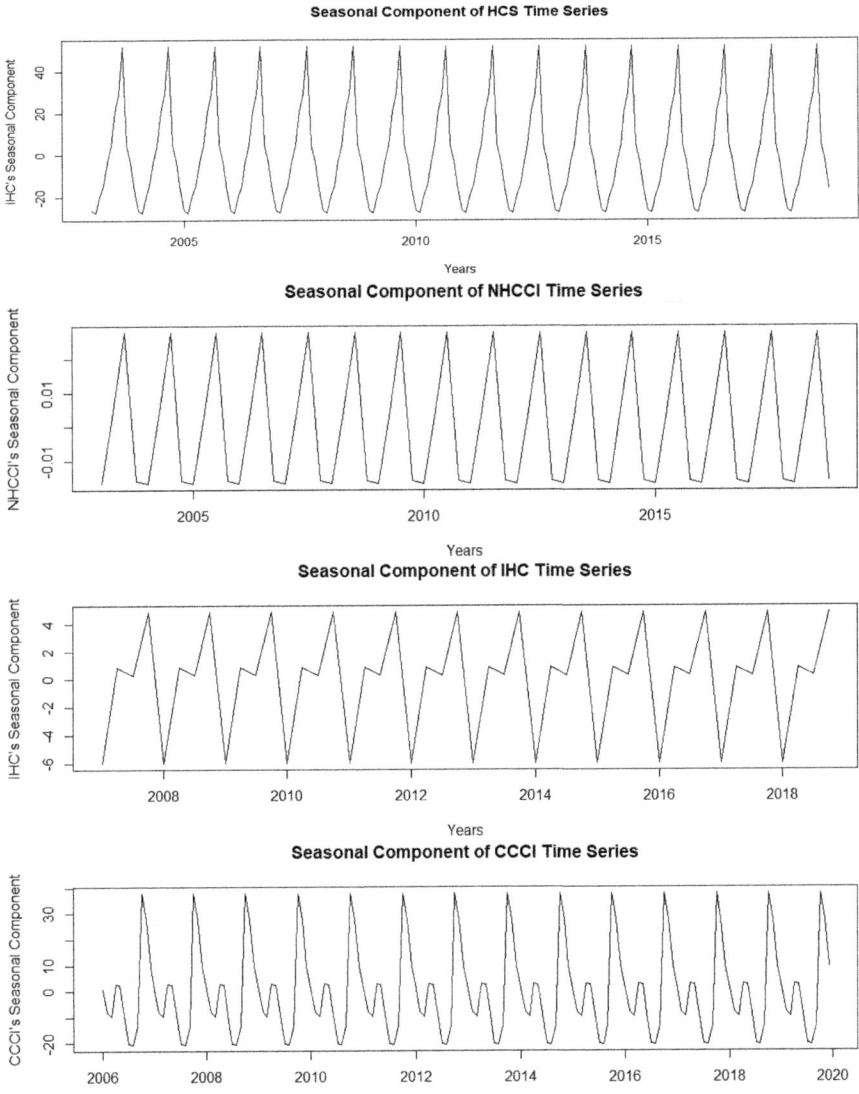

Fig. 2.4 The seasonal components of HCS, NHCCI, IHC, and CCCI time series

- Transforming the time series to acquire a stationary time series in case of a non-stationary time series
- Choosing an appropriate model to fit the stationary time series
- Using the trained model for forecasting the time series
- Conducting diagnostic tests (i.e., residual analysis)

R Code 2.5
Time series decomposition and seasonal component extraction

```
hcs.decomposed <- decompose(hcs.ts)
nhcci.decomposed <- decompose(nhcci.ts)
ihc.decomposed <- decompose(ihc.ts)
ccci.decomposed <- decompose(ccci.ts)
plot(hcs.decomposed$seasonal, main = "Seasonal Component of HCS Time Series",
xlab = "Years", ylab = "IHC's Seasonal Component")
plot(nhcci.decomposed$seasonal, main = "Seasonal Component of NHCCI Time Series"
, xlab = "Years", ylab = "NHCCI's Seasonal Component")
plot(ihc.decomposed$seasonal, main = "Seasonal Component of IHC Time Series",
xlab = "Years", ylab = "IHC's Seasonal Component")
plot(ccci.decomposed$seasonal, main = "Seasonal Component of CCCI Time Series",
xlab = "Years", ylab = "CCCI's Seasonal Component")
```

2.4.1 Moving Average (MA)

A moving average (MA) model is the simplest yet powerful method for smoothing time series. It removes or reduces random fluctuations in a time series. A MA model is represented as a linear regression of the current value of a series against a series following a white noise process. So, these white noises are propagated into the upcoming values of the time series. A MA model is expressed as follows:

$$X_t = \mu + A_t - \theta_1 A_{t-1} - \theta_2 A_{t-2} - \theta_3 A_{t-3} - q \tag{2.3}$$

where X_t is the value of time series at time t, μ is the mean of the series, A_{t-i} is a white noise term, $\theta_1,\ldots,\theta_q$ are the model parameters, and q is the order of the MA model. ACF and PACF plots of a stationary time series are used to find the order of the moving average model. If the ACF plot shows significant spikes through first lags till the lag p and the PACF plot shows a geometric decay, the order of the moving average model is p.

A first-order MA model is used to fit the NHCCI time series from 2003 to 2017 using R Code 2.7. Table 2.3 shows the NHCCI forecast for the four quarters of 2018. Figure 2.6 illustrates the NHCCI time series from 2003 to 2017 and its forecasts for the four quarters of 2018.

2.4.2 Autoregressive (AR)

An autoregressive (AR) model uses previous values of a time series as the inputs in a linear regression model to predict future values in the time series. Therefore, a time series' predictions are linearly dependent on the past values of the time series and a white noise process. An AR model is expressed by:

2.4 Univariate Time Series Forecasting

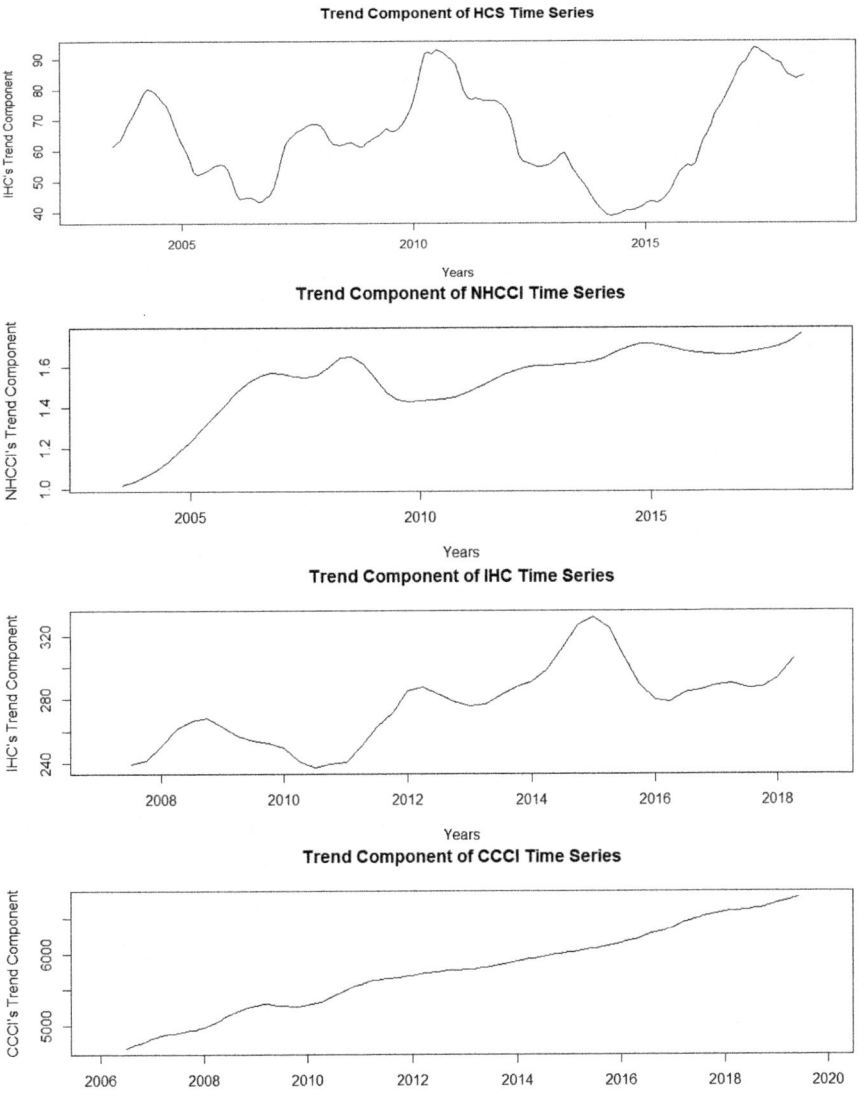

Fig. 2.5 Trend components of HCS, NHCCI, IHC, and CCCI time series

$$X_t = \delta + \phi_1 X_{t-1} + \phi_2 X_{t-2} + \cdots + \phi_p X_{p-1} + A_t \quad (2.4)$$

where X_t is the time series' value at time t, A_t is white noise at time t, $\phi_1,\ldots, \phi_p$ are model parameters, and p is the order of the AR model. δ is expressed by:

$$\delta = \left(1 - \sum_{p}^{i=1} \phi_i\right) \mu \quad (2.5)$$

R Code 2.6
Trend component extraction of time series

```
plot(hcs.decomposed$trend, main = "Trend Component of HCS Time Series" , xlab =
"Years", ylab = "IHC's Trend Component")
plot(nhcci.decomposed$trend, main = "Trend Component of NHCCI Time Series" , xlab
= "Years", ylab = "NHCCI's Trend Component")
plot(ihc.decomposed$trend, main = "Trend Component of IHC Time Series" , xlab =
"Years", ylab = "IHC's Trend Component")
plot(ccci.decomposed$trend, main = "Trend Component of CCCI Time Series" , xlab
= "Years", ylab = "CCCI's Trend Component")
```

R Code 2.7
Moving average forecasting model (first order)

```
library(forecast)
nhcci_train.ts <- ts(nhcci, frequency = 4, start = c(2003,1), end =c(2017,4) )
ma1_model<- ma(nhcci_train.ts, order= 1)
ma1_forecast<- forecast(ma1_model, 4)
ma1_forecast
plot(ma1_forecast, main = "NHCCI Forecast Using A First-Order Moving Average
Model", xlab = "Year", ylab = "NHCCI")
```

Table 2.3 NHCCI time series forecasts for the four quarters of 2018 with 80% and 95% confidence intervals using a first-order MA model

Point	Forecast	Lo 80	Hi 80	Lo 95	Hi 95
2018-Q1	1.667918	1.602178	1.733659	1.567377	1.76846
2018-Q2	1.688462	1.584911	1.792012	1.530095	1.846828
2018-Q3	1.715111	1.57433	1.855891	1.499805	1.930416
2018-Q4	1.668647	1.496876	1.840418	1.405947	1.931348

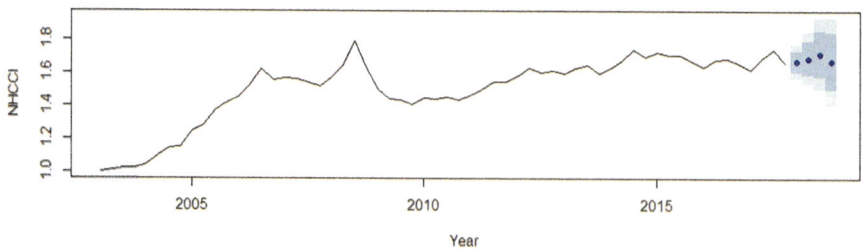

Fig. 2.6 NHCCI time series forecasts for the four quarters of 2018 with 80% and 95% confidence intervals using a first-order MA model

2.4 Univariate Time Series Forecasting

where μ is the process mean and p defines the order of an AR model. For example, an AR(1) autoregressive model is called a first-order process. The predicted value by a first-order process is determined based on the previous value, while an AR(2) process determines the current value based on the two preceding values. Akaike information criterion (AIC) is often used to select the optimum order for an AR model. Ordinary least squares ("OLS"), "Yule-Walker," and "burg" approaches can be used to fit the AR model.

An AR model is developed to forecast CCCI for 2018. ACF and PACF plots are used to determine an appropriate order for an AR model. Figure 2.7 illustrates the ACF and PACF plots for the CCCI time series. This figure shows a geometric decay of significant autocorrelations for many lags in the CCCI time series. Conversely, the PACF plot demonstrates a significant spike only at lag 1. It means all the higher-order autocorrelations are well explained by the lag-1 autocorrelation. These two plots suggest a first-order AR model for CCCI.

R Code 2.8 is used to model the first-order AR to forecast CCCI for 1 year. Table 2.4 shows the CCCI forecast for 12 months of 2018.

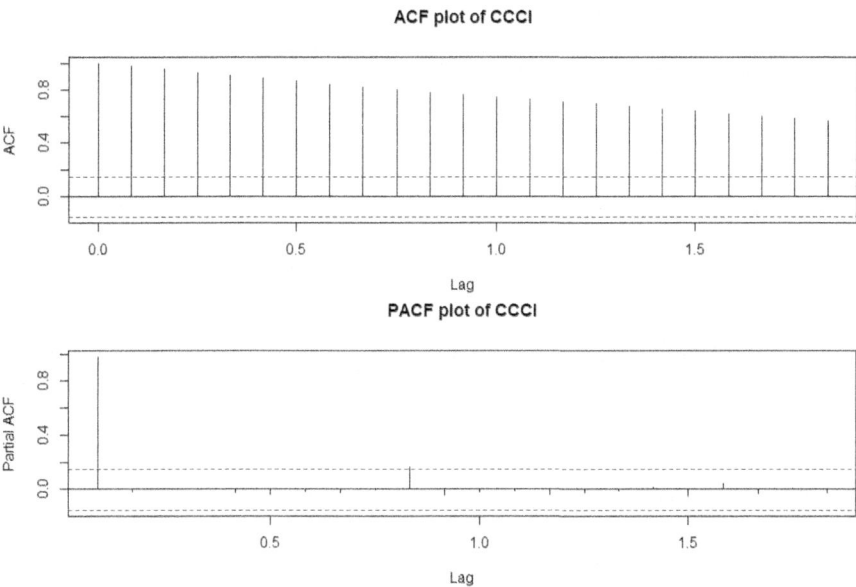

Fig. 2.7 ACF and PACF plots of CCCI time series

R Code 2.8
Autoregressive forecasting model (first order)

```
library("forecast")
ccci_train.ts <- ts(ccci, frequency = 12, start = c(2006,1), end =c(2017,12) )
ar1_model<- ar(ccci_train.ts, aic= TRUE, method= "ols", order.max = 12)
ar1_forecast<- forecast(ar1_model, 12)
plot(ar1_forecast, main = "CCCI Forecast Using A First Order Autoregressive Model", xlab = "Year", ylab = "CCCI")
```

Table 2.4 CCCI time series forecasts for 2018 with 80% and 95% confidence intervals using a first-order AR model

Point	Forecast	Lo 80	Hi 80	Lo 95	Hi 95
Jan 2018	6611.183	6564.383	6657.983	6539.609	6682.758
Feb 2018	6624.532	6559.52	6689.544	6525.104	6723.96
Mar 2018	6634.252	6558.041	6710.462	6517.698	6750.806
Apr 2018	6637.813	6553.427	6722.199	6508.756	6766.87
May 2018	6644.301	6552.077	6736.525	6503.257	6785.345
Jun 2018	6664.13	6564.261	6764.000	6511.394	6816.867
Jul 2018	6682.946	6576.407	6789.486	6520.008	6845.885
Aug 2018	6697.622	6585.425	6809.819	6526.031	6869.213
Sep 2018	6710.257	6593.75	6826.765	6532.075	6888.44
Oct 2018	6722.883	6602.886	6842.879	6539.364	6906.402
Nov 2018	6736.776	6613.058	6860.493	6547.566	6925.986
Dec 2018	6751.101	6623.482	6878.721	6555.924	6946.279

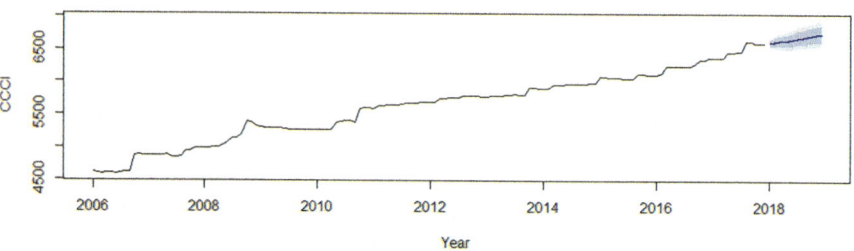

Fig. 2.8 CCCI time series forecast for 2018 with 80% and 95% confidence intervals using a first-order AR model

Figure 2.8 illustrates the CCCI time series from 2003 to 2017 and its forecasts for 2018.

2.4.3 Exponential Smoothing (ES)

Exponential smoothing is a simple time series model that is mainly used for short-range forecasting. The exponential smoothing forecasting method uses an exponentially decreasing weight for past observations to predict future values. The three types of exponential smoothing forecasting methods are single (known as simple), double (known as Holt), and triple (known as Holt-Winters) exponential smoothing models. Single exponential smoothing is used to forecast a univariate time series with no trend or seasonality. Double exponential smoothing extends the single

2.4 Univariate Time Series Forecasting

exponential smoothing that captures trends in univariate time series modeling. Holt-Winters or triple exponential smoothing is the extended version of double exponential smoothing that captures seasonality as well.

A single exponential smoothing model is expressed by:

$$S_{t+1} = \alpha y_t + (1-\alpha) * S_t, \qquad 0\alpha \leq 1, t0. \qquad (2.6)$$

This formulation can also be expressed by:

$$S_{t+1} = S_t + \alpha \varepsilon_t \qquad (2.7)$$

where ϵ_t is the model forecast error for period t.

A Holt (i.e., double exponential smoothing) is an extension of single exponential smoothing to capture trends; it is expressed by:

$$S_t = \alpha yt + (1-\alpha) * (S_{t-1} + b_{t-1}), \qquad 0 \leq \alpha \leq 1 \qquad (2.8)$$

$$b_t = \gamma * (S_t - S_{t-1}) + (1-\gamma) * b_{t-1} \qquad 0 \leq \gamma \leq 1 \qquad (2.9)$$

For the initial value of b_1, these three suggestions can be used (Heckert et al., 2002):

$$b_1 = y_2 - y_1 \qquad (2.10)$$

$$b_1 = 1/3 * \left[(y_2 - y_1) + (y_3 - y_2) + (y_4 - y_3) \right] \qquad (2.11)$$

$$b_1 = (y_n - y_1)/(n-1) \qquad (2.12)$$

The double exponential smoothing forecasting model for m period ahead is expressed by:

$$F_{t+m} = S_t + mb_t \qquad (2.13)$$

A triple exponential smoothing model is expressed by:

$$S_t = \alpha yt + (1-\alpha) * (S_{t-1} - 1 + b_{t-1}) \, 0 \leq \alpha \leq 1 \qquad (2.14)$$

$$b_t = \gamma * (S_t - S_{t-1}) + (1-\gamma) * b_{t-1} \, 0 \leq \gamma \leq 1 \qquad (2.15)$$

$$I_t = \beta * (y_t / S_t) + (1-\beta) * I_{t-L} \, 0 \leq \beta \leq 1 \qquad (2.16)$$

The triple exponential smoothing forecasting model for m period ahead is expressed by:

$$F_{t+m} = (S_t + mb_t) * I_{t-L+m} \qquad (2.17)$$

where y_t is the observation, S_t is the smoothed observation, b_t is the trend factor, I_t is the seasonal index, F_{t+m} is the forecast at m periods ahead, and t is the time index.

Single, double, and triple exponential smoothing models are developed to forecast IHC for 2018. R Code 2.9 is used to create all three exponential smoothing models to forecast IHC for 1 year ahead (four quarters). Table 2.5 shows the IHC forecasts for the four quarters of 2018 using single, double, and triple exponential smoothing models.

Figure 2.9 illustrates the IHC time series from 2003 to 2017 and 2018 forecasts.

R Code 2.9
Single, double, and triple exponential smoothing models

```
library(forecast)
ihc_train.ts <- ts(ihc, frequency = 4, start = c(2007,1), end =c(2017,4) )
ses.ihc<- ses(ihc_train.ts, alpha = .5, h = 4)
plot(ses.ihc, main = "IHC Forecast Using A Single Exponential Smoothing", xlab =
"Year", ylab = "IHC")
ses.ihc
des.ihc <- holt(ihc_train.ts,h = 4)
plot(des.ihc, main = "IHC Forecast Using A Double Exponential Smoothing (Holt)",
xlab = "Year", ylab = "IHC")
des.ihc
HW.ihc <- hw(ihc_train.ts,h = 4)
plot(HW.ihc, main = "IHC Forecast Using A Triple Exponential Smoothing (Holt-
Winter)", xlab = "Year", ylab = "IHC")
HW.ihc
```

Table 2.5 IHC time series forecasts for 2018 using single, double, and triple exponential smoothing models

Point	Single	Double	Triple
2018-Q1	284.0437	286.1209	274.4531
2018-Q2	284.0437	287.3298	287.1468
2018-Q3	284.0437	288.5387	284.3694
2018-Q4	284.0437	289.7476	290.1898

2.4 Univariate Time Series Forecasting

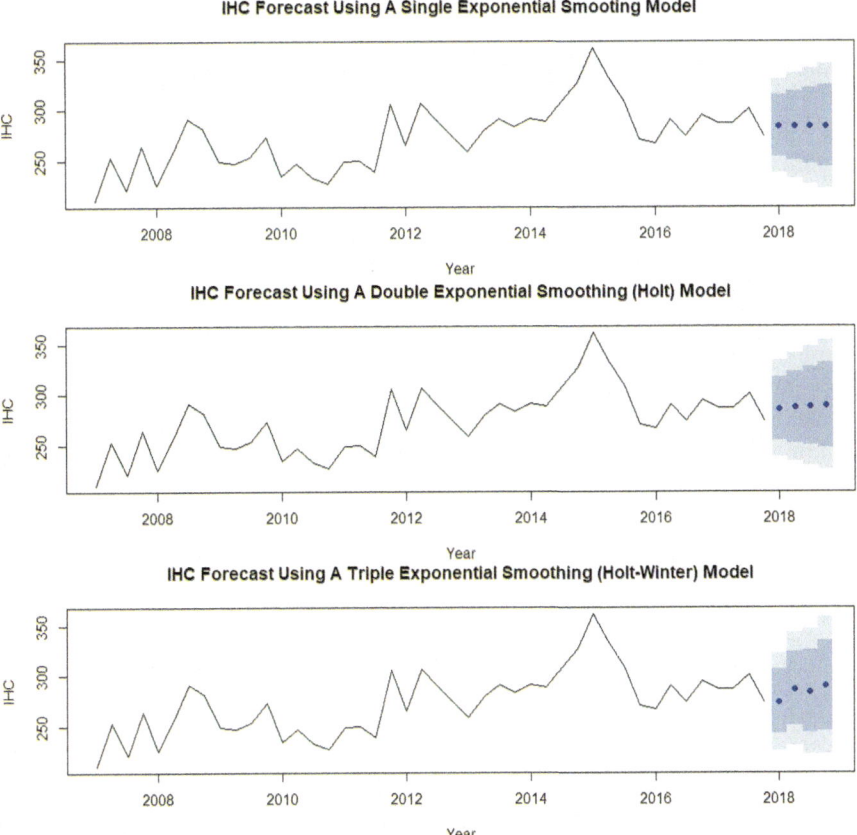

Fig. 2.9 IHC time series forecasts for 2018 with 80% and 95% confidence intervals using single, double, and triple exponential smoothing models

2.4.4 Autoregressive Moving Average (ARMA)

Box and Jenkins popularized an approach that uses both moving average and autoregressive processes to model a time series (Ho et al., 2002). This method is known as ARMA or Box-Jenkins. The main underlying assumption of an ARMA model is the stationarity of the time series. Therefore, the nonstationary data needs to be transformed before an ARMA model is created. X_t is an ARMA(p, q) process if it is stationary and if for every t,

$$X_t - \varphi_1 X_{t-1} - \cdots - \varphi_p X_{t-p} = Z_t + \theta_1 Z_{t-1} + \cdots + \theta_q Z_{t-q} \tag{2.18}$$

where Z_t is white noise (0, σ^2), φ is the autoregressive model's parameter, and θ is the moving average parameter.

Autocorrelation function (ACF) and partial autocorrelation function (PACF) can help discover AR and MA orders of a stationary time series. Watson and Teelucksingh (2002) suggested graphical rules to select appropriate AR and MA orders based on the ACF and PACF plot. ACF and PACF values of white noise are equal to zero at all lags. A slow decay in ACF values of a time series with significant spikes in PACF values indicates an AR process, while significant spikes in ACF values of a time series with a slow decay in PACF values show a MA process.

A time series shows an "AR signature" if a sharp cutoff is shown in the PACF plot, whereas the ACF decays more slowly. It means that the autocorrelation in the time series can be modeled by adding AR terms than MA terms. On the other hand, a time series displays an "MA signature" if the ACF plot shows a sharp drop, while the PACF decays more slowly. The autocorrelation can be explained by adding MA terms rather than AR terms. If both ACF and PACF values show a slow decay, the time series follows an ARMA process (Nau, 2017). Since IHC has weak stationarity according to the ACF test results shown in Fig. 2.2, an autoregressive moving average (ARMA) model is used to forecast IHC for 1 year (i.e., four quarters) without differencing the time series. Figure 2.10 shows both ACF and PACF plots of the IHC time series. The ACF plot shows a decaying pattern before lag 5, suggesting a MA process with order 4. Conversely,

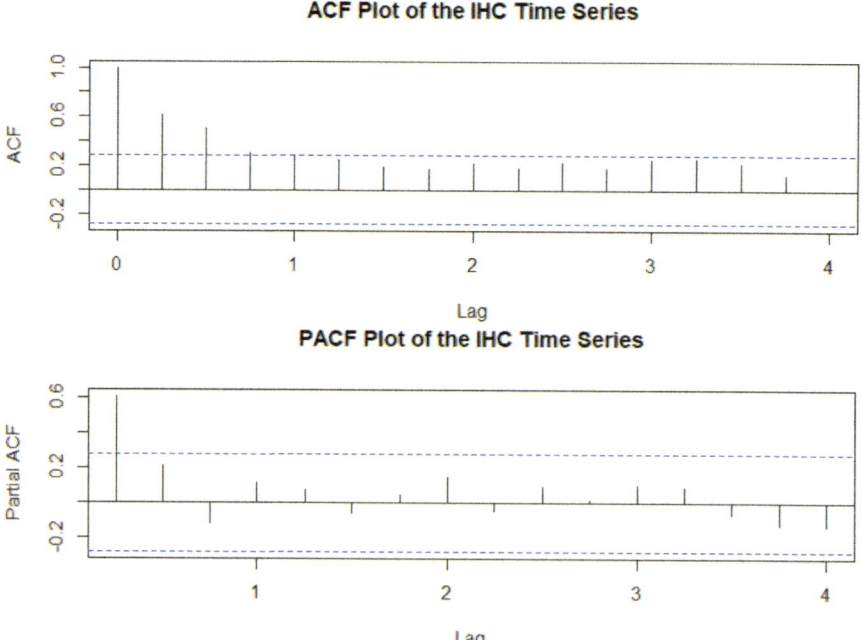

Fig. 2.10 ACF and PACF plots of IHC time series

2.4 Univariate Time Series Forecasting

Table 2.6 The forecasted values for the simulated time series with 80% and 95% confidence intervals using an ARMA(1,1) model

Point	Point forecast	Lo 80	Hi 80	Lo 95	Hi 95
2018-Q1	266.9771	234.0169	299.9372	216.5688	317.3853
2018-Q2	267.0851	230.6477	303.5226	211.3589	322.8114
2018-Q3	264.2654	224.8156	303.7151	203.9322	324.5985
2018-Q4	265.7154	226.2404	305.1904	205.3436	326.0873

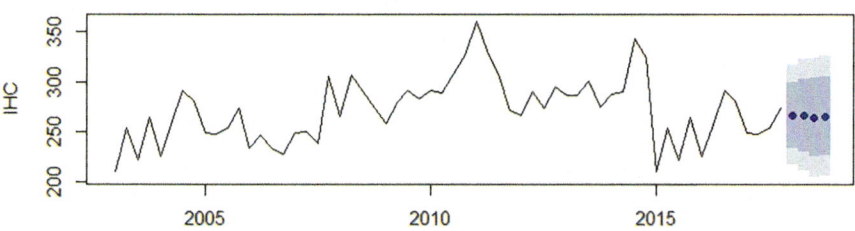

Fig. 2.11 The forecasted values for the IHC time series with 80% and 95% confidence intervals using an ARMA(1,5) model

the PACF plot illustrates just one significant spike at lag 1, indicating an AR process with order 1.

R Code 2.10 is used to develop an ARMA(1,4) model to forecast the HCS time series. Table 2.6 shows forecasted values for the ARMA(1,1) model. Figure 2.11 illustrates the forecasted values of the IHC time series using the ARMA(1,1) model.

R Code 2.10
ARMA(1,1) model

```
ihc_train.ts <- ts(ihc, frequency = 4, start = c(2003,1), end =c(2017,4) )
library("forecast")
arma11<- arima(ihc_train.ts, order=c(1,0,5))
arma11_forecast<- forecast(arma11, 4)
plot(arma11_forecast, main = "Forecast of IHC Using An ARMA (1,5) Model", ylab =
"IHC")
```

2.4.5 Autoregressive Integrated Moving Average (ARIMA)

The autoregressive integrated moving average (ARIMA) is the generalized version of the ARMA model that can handle the stationarity of a time series by adding a differencing parameter. An ARIMA model has three parameters p, d, and q. The order (the number of time lags) of the autoregressive model, the order of the moving average, and the degree of differencing (the number of times the data is differenced) are denoted by p, q, and d, respectively. X_t is an ARIMA(p, d, q) process if $Y_t = (1 - B)d\, X_t$ is a causal ARMA(p, q) process. Equation 2.19 expresses the generalized formulation of an ARIMA model.

$$\varphi*(B)X_t \equiv \varphi(B)(1-B)^d X_t = \theta(B)Z_t \qquad (2.19)$$

where Z_t represents white noise; $\varphi(z)$ and $\theta(z)$ are polynomials of degrees p and q, respectively; and $\varphi(z)$ is not equal to 0 for $|z|$ greater or equal to 1.

An ARIMA model is used to forecast NHCCI for 1 year. Based on the ADF test results, two degrees of differencing make the NHCCI stationary. Therefore, the differencing parameter of the ARIMA model is set to 2. To approximate the order of AR and MA components of the ARIMA model, ACF and PACF plots are used. Figure 2.12 illustrates both ACF and PACF values of the NHCCI training dataset. The ACF plot of the NHCCI training dataset shows a geometric decay. On the other hand, the PACF shows a significant spike at the first lag. These results suggest that AR and MA orders are 1 and 0, respectively. Therefore, the ARIMA model for NHCCI can be expressed by ARIMA(1,2,0).

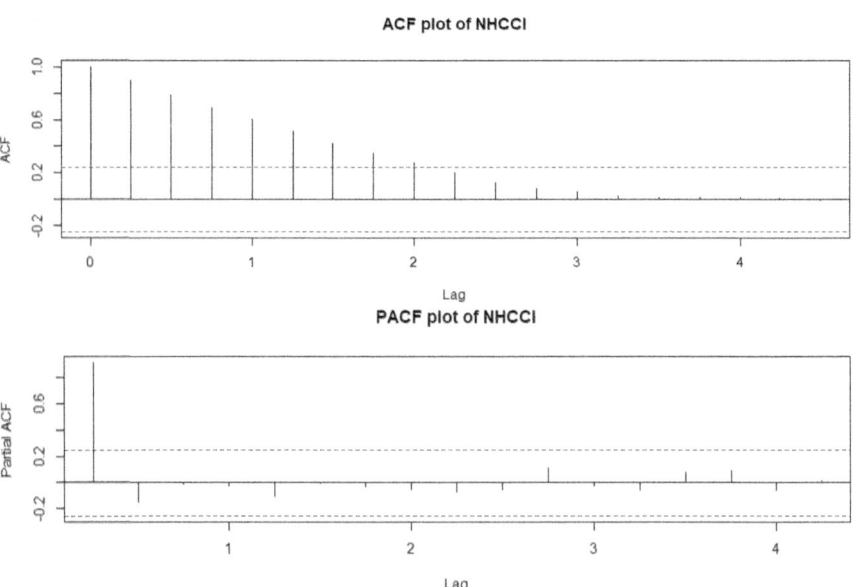

Fig. 2.12 ACF and PACF plots of NHCCI time series

2.4 Univariate Time Series Forecasting

Table 2.7 NHCCI time series forecasts for 2018 with 80% and 95% confidence intervals using ARIMA(1,2,0)

Point	Forecast	Lo 80	Hi 80	Lo 95	Hi 95
2018-Q1	1.656321	1.577005	1.735636	1.535017	1.777624
2018-Q2	1.649351	1.497373	1.801329	1.416921	1.881781
2018-Q3	1.638220	1.394477	1.881962	1.265448	2.010991
2018-Q4	1.628610	1.281163	1.976056	1.097235	2.159984

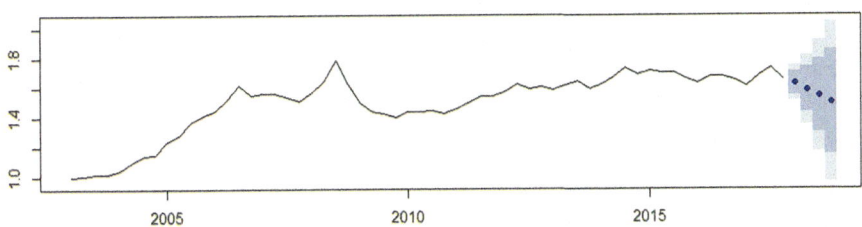

Fig. 2.13 NHCCI time series forecast for 2018 with 80% and 95% confidence intervals using ARIMA(1,2,0)

R Code 2.11
Autoregressive integrated moving average (ARIMA)

```
arima_nhcci <- arima(nhcci_train.ts, order = c(1,2,0))
arima_forecast = forecast(arima_nhcci, 4)
arima_forecast
plot(arima_forecast , main = "NHCCI time series forecast Using an ARIMA(1,2,0) Model")
```

R Code 2.11 generates the ARIMA model of NHCCI and forecasts NHCCI for 1 year (four questers). Table 2.7 shows the NHCCI forecasts for the four quarters of 2018 using the model. Figure 2.13 illustrates the prediction of NHCCI time series for 1 year using ARIMA(1,2,0).

2.4.6 Seasonal Autoregressive Integrated Moving Average (SARIMA)

One of the drawbacks of ARIMA models is that they do not consider the seasonality in seasonal time series data. Seasonal time series have a repeating cycle over time. A seasonal autoregressive integrated moving average (SARIMA) is the extension of an ARIMA model that explicitly captures a seasonal behavior in time series data. A seasonal ARIMA model has four additional seasonal parameters (e.g., P, D, Q, and

s). These parameters are very similar to the nonseasonal components of the ARIMA model (i.e., p, d, and q), but they involve modeling the seasonal components in a time series (Hyndman & Athanasopoulos, 2018). If d and D are nonnegative integers, then X_t is a seasonal ARIMA(p, d, q) × (P, D, Q)s process with period s if the differenced series $Y_t = (1 - B)^d (1 - B^s)^D X_t$ is a causal ARIMA process expressed by:

$$\varphi(B)\Phi(B^s)Y_t = \theta(B)\Theta(B^s)Z_t \qquad (2.20)$$

where Z_t is white noise, S is the number of observations per season, and $\varphi(z)$, $\Phi(z)$, $\theta(z)$, and $\Theta(z)$ are polynomials of degrees p, P, q, and Q, respectively.

"Parsimony principle" helps determine the parameters of a SARIMA model. Parsimony or economy of parameters is a desirable criterion suggested by Tukey (1961) and Box and Jenkins (1976). This principle indicates that the simplest possible model should be chosen among the candidate models. Therefore, models with the least Akaike information criterion (AIC) or Bayesian information criterion (BIC) are selected as the simplest SARIMA model.

A seasonal autoregressive integrated moving average model is selected for forecasting Highway Construction Spending (HCS) since it is highly seasonal. Therefore, seven parameters must be determined for the SARIMA model (e.g., p, d, q, P, D, Q, and s). The seasonality period (i.e., s) is set to 12. Further, d is equal to 1 due to the non-stationarity of the HCS time series (see the result of the ADF test for HCS in Table 2.1). Moreover, this time series is needed to be seasonally differenced once due to its seasonality. This could be easily found using the ACF and PACF plots of a time series. Seasonal differences may be needed in the model if the pattern of the ACF repeats itself on the seasonal frequencies and spikes exist in the PACF at the seasonal frequency. Based on the ACF and PACF plots of HCS (Fig. 2.14), the time series needs to be differenced once. Therefore, the initial SARIMA model will be in the form of SARIMA(p,1,q)(P,1,Q)$_{12}$.

The initial SARIMA model (i.e., SARIMA(0,1,0)(0,1,0)$_{12}$) can be generated using R Code 2.12.

R Code 2.12
SARIMA(0,1,0)(0,1,0)$_{12}$

```
library(astsa)
hcs_train <- ts(hcs, frequency = 12, start = c(2003,1), end =c (2017,12))
Sarima_010010_HCS <- sarima(hcs_train, 0,1,0,0,1,0,12)
```

Figure 2.15 illustrates the summary statistics for the residuals of the generated model.

Figure 2.15 displays significant p-values for the Ljung-Box test for all of the residuals, indicating that there is autocorrelation between the residuals. The ACF plot of the initial model's residuals shows significant spikes through the first and

2.4 Univariate Time Series Forecasting

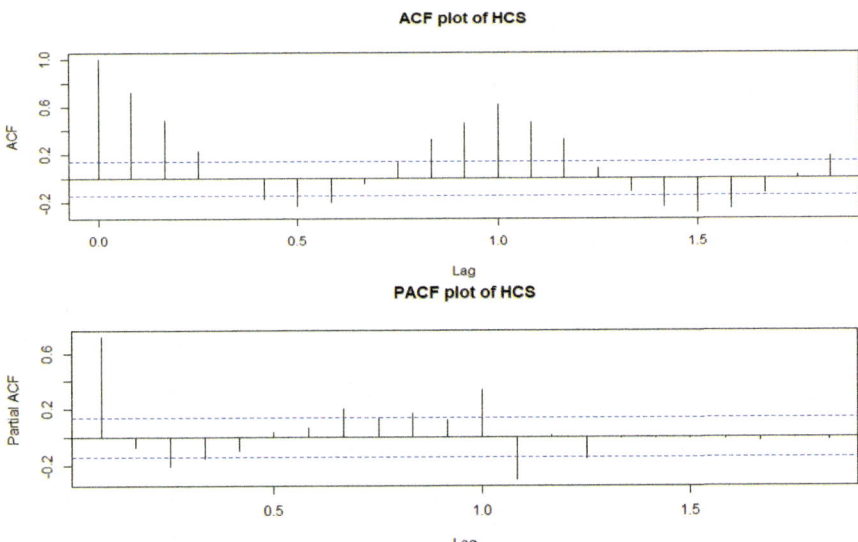

Fig. 2.14 ACF and PACF plots of HCS time series

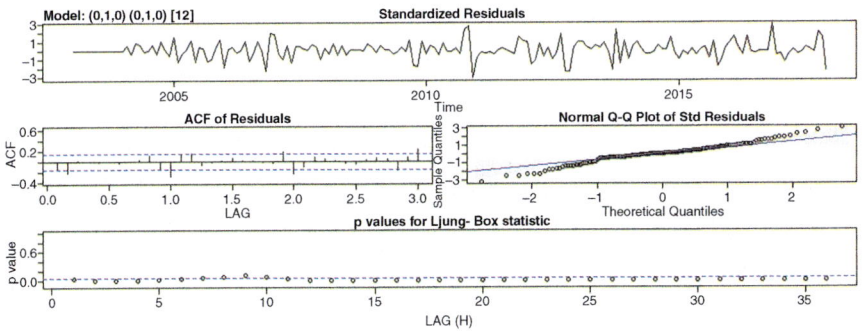

Fig. 2.15 The summary of SARIMA$(0,1,0)(0,1,0)_{12}$ model

other seasons, indicating the existence of nonseasonal and seasonal MA processes. One and two are the potential nonseasonal orders for the MA process. We also consider three orders for the seasonal MA process. The first potential SARIMA model is SARIMA$(0,1,1)(0,1,1)12$. R Code 2.13 is used to develop this model

R Code 2.13
SARIMA$(0,1,1)(0,1,1)_{12}$

```
Sarima_011011_HCS <- sarima(hcs_train, 0,1,1,0,1,1,12)
```

Figure 2.16 illustrates the summary statistics for the residuals of the SARIMA(0,1,1)(0,1,1)$_{12}$ model. The results of the Ljung-Box test show significant *p*-values for the residuals, indicating that there is still autocorrelation among the residuals. Moreover, the ACF plot of the residuals confirms the results of the Ljung-Box test by displaying significant spikes in the residuals at lags 2 and 24. Negative autocorrelations at lag s indicate a seasonal MA process (Nau, 2017).

One more seasonal and nonseasonal MA order is added to the previous model to remove the significant spikes. Therefore, a SARIMA(0,1,2)(0,1,2)$_{12}$ is developed using R Code 2.14.

Figure 2.17 shows the summary statistics for the residuals of the model. The results of the Ljung-Box test show no significant *p*-values for the residuals of the

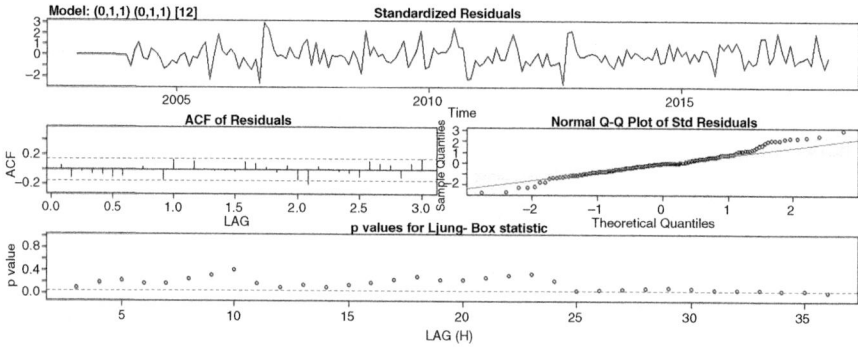

Fig. 2.16 The summary of SARIMA(0,1,1)(0,1,1)$_{12}$ model

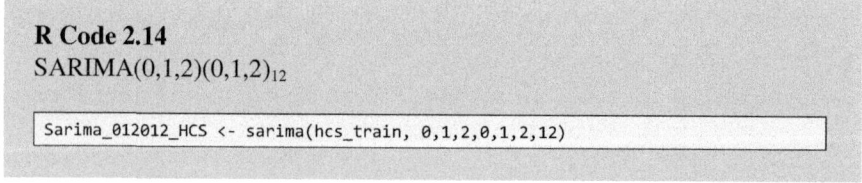

R Code 2.14
SARIMA(0,1,2)(0,1,2)$_{12}$

```
Sarima_012012_HCS <- sarima(hcs_train, 0,1,2,0,1,2,12)
```

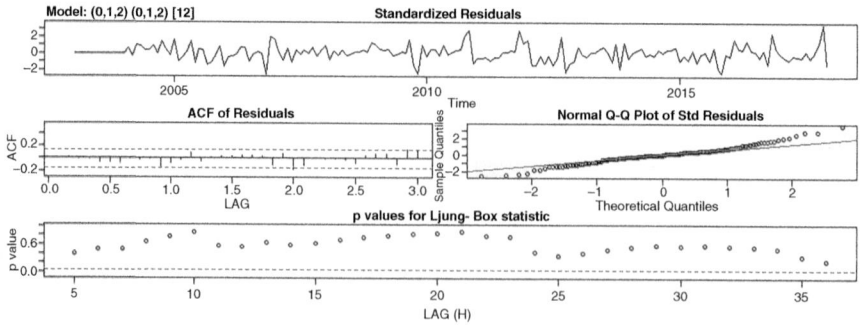

Fig. 2.17 The SARIMA(0,1,2)(0,1,2)$_{12}$

2.4 Univariate Time Series Forecasting

model. However, the ACF plot still shows a significant spike in the second season, indicating that there is still autocorrelation between the residuals. This plot suggests adding one order to the seasonal MA process.

R Code 2.15 is used to develop SARIMA$(0,1,2)(0,1,3)_{12}$ model.

Figure 2.18 shows the summary statistics for the residuals of the model. The results of the Ljung-Box test show no significant p-values for the residuals of the model, meaning that the model fits the data correctly. Also, the ACF plot confirms the results of the Ljung-Box test, indicating no autocorrelation among the residuals.

Table 2.8 shows the AIC and BIC values for three SARIMA models developed in this section. The AIC and BIC values also confirm that the SARIMA(0,1,2)

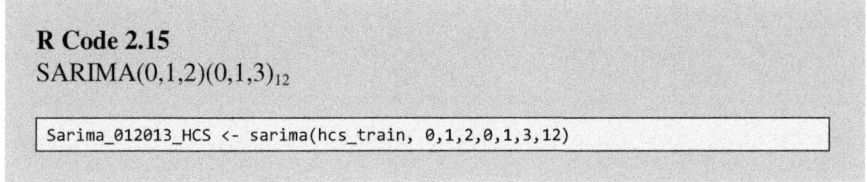

R Code 2.15
SARIMA$(0,1,2)(0,1,3)_{12}$

```
Sarima_012013_HCS <- sarima(hcs_train, 0,1,2,0,1,3,12)
```

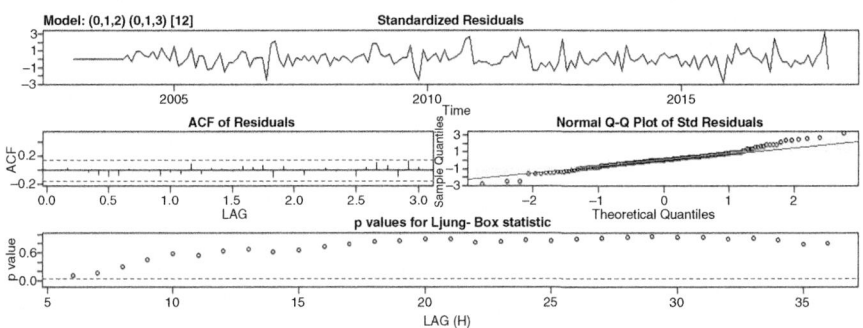

Fig. 2.18 The summary of SARIMA$(0,1,2)(0,1,3)_{12}$ model

Table 2.8 AIC and BIC values for all SARIMA models

Models	AIC	BIC
SARIMA$(0,1,1)(0,1,1)_{12}$	7.997032	8.049583
SARIMA$(0,1,2)(0,1,2)_{12}$	7.93911	8.026694
SARIMA$(0,1,2)(0,1,3)_{12}$	7.899997	8.005098

$(0,1,3)_{12}$ model is recommended (parsimony principle). Therefore, SARIMA$(0,1,2)$ $(0,1,3)_{12}$ is used to forecast HCS for 1 year (12 months).

R Code 2.16 is used to forecast HCS for 12 months using SARIMA$(0,1,2)$ $(0,1,3)_{12}$. Table 2.9 shows the forecasts for HCS for 12 months.

Figure 2.19 shows the HCS forecasts with 80% and 90% confidence intervals for 12 months using SARIMA$(0,1,2)(0,1,3)_{12}$.

R Code 2.16
HCS forecast using SARIMA$(0,1,2)(0,1,3)_{12}$

```
sarima_forecast <- sarima.for(hcs_train, 12, 0, 1, 2, P = 0, D = 1, Q = 3, S = 12)
sarima_forecast
```

Table 2.9 HCS forecasts with standard errors for 12 months using SARIMA$(0,1,2)(0,1,3)_{12}$ model

Point	Forecast	Standard error
Jan 2018	64.64307	14.84001
Feb 2018	57.89494	17.71735
Mar 2018	68.96394	18.97541
Apr 2018	77.29724	20.15509
May 2018	86.63034	21.26945
Jun 2018	99.85132	22.32825
Jul 2018	117.51224	23.33908
Aug 2018	112.46465	24.3079
Sep 2018	135.29342	25.23957
Oct 2018	98.00904	26.13804
Nov 2018	81.95908	27.00665
Dec 2018	63.9463	27.84853

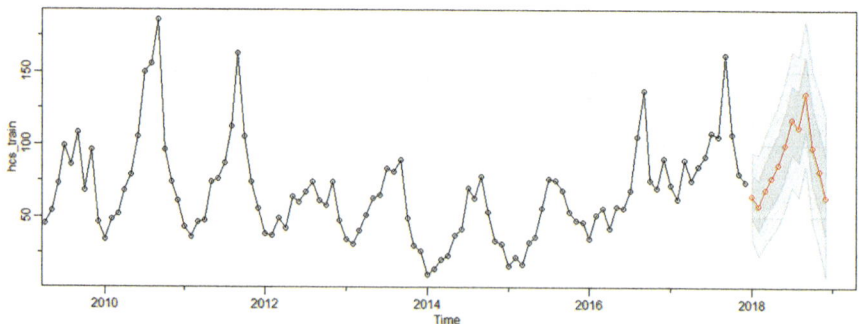

Fig. 2.19 HCS forecasts with 80% and 90% confidence intervals for 12 months using a SARIMA$(0,1,2)(0,1,3)_{12}$ model

2.5 Diagnostic Tests for Time Series Models

The underlying assumptions of all the univariate time series models discoursed in the previous sections are the stationarity and linearity of the time series. Diagnostic tests examine whether the modeling assumptions are satisfied, and the residuals of a time series model mimic a white noise process, which indicates the time series model has fully captured the information of the time series. A robust univariate time series model with a good fit satisfies the following assumptions for model residuals:

- The residuals are uncorrelated. If there is a correlation between the residuals, then information remains in the residuals that can improve forecasting accuracy.
- The residuals have a constant variance. When the residuals have variable variance, known as heteroscedasticity, the stationarity assumption of a time series does not hold.
- The residuals should follow a normal distribution. If the model residuals are normally distributed, the standardized forecast errors, confidence intervals, and p-values are estimated unbiasedly. However, in large sample sizes, violations of this normality assumption often do not noticeably impact results (Schmidt & Finan, 2018).

Different diagnostic tests and graphical illustrations are suggested to diagnose these three modeling assumptions. Ljung-Box test and autocorrelation function (ACF) are commonly used to check for no autocorrelation among model residuals. The ARCH-LM (autoregressive conditional heteroscedasticity-Lagrange multiplier) test and the ACF plot of squared residuals can be applied to examine the assumption of constant variance. Shapiro-Wilk test and the normal quantile-quantile (Q-Q) plot are commonly used to diagnose normality in model residuals.

2.5.1 Diagnostic Tests for No Autocorrelation

Two diagnostic tests are conducted to check that there is no autocorrelation among model residuals.

2.5.1.1 Ljung-Box Test

Ljung-Box test is one of the popular statistical hypothesis tests for diagnosing the existence of serial autocorrelation among the model residuals. The null hypothesis of this test is that the model residuals are not autocorrelated. The rejection of the null hypothesis for this test means that the residuals are autocorrelated. If the time series model satisfies the assumption of no autocorrelation between the residuals at lag m, the Ljung-Box statistic Q follows asymptotically a chi-squared distribution (with $m - g$ degrees of freedom, where g denotes the number of parameters in the model). R Code 2.17 is used to test the absence of autocorrelation between the residuals of SARIMA(0,1,2)(0,1,3)$_{12}$, which is developed for HCS forecasting in the previous section using the Ljung-Box test.

Table 2.10 shows the results for the Ljung-Box test for the SARIMA(0,1,2) (0,1,3)$_{12}$ model. Based on the results, there is not enough evidence to reject the null hypothesis of the Ljung-Box test (no autocorrelation) for the residuals of the SARIMA(0,1,2)(0,1,3)$_{12}$ model at the 5% level of significance. Therefore, the SARIMA(0,1,2)(0,1,3)$_{12}$ model satisfies the first assumption of no autocorrelation between the residuals.

R Code 2.17
Ljung-Box test
```
resid <-residuals(Sarima_012013_HCS$fit)
Box.test(resid,type = "Ljung-Box")
```

2.5.1.2 Autocorrelation Function

The autocorrelation function (ACF) demonstrates the autocorrelations of a time series for different lags. Since time series models assume the model residuals as white noise, the autocorrelations between the residuals are supposed to be insignificant in the ACF. R Code 2.18 is used to illustrate the ACF plot to diagnose no autocorrelation assumption between the residuals of the SARIMA(0,1,2)(0,1,3)$_{12}$.

Figure 2.20 presents the ACF of the residuals for SARIMA(0,1,2)(0,1,3)$_{12}$. No significant autocorrelation can be found between the model residuals in Fig. 2.20.

Table 2.10 Result of the Ljung-Box test for the residuals of the SARIMA(0,1,2)(0,1,3)$_{12}$ model

Chi-squared	Degree of freedom	p-value
0.016	1	0.89

R Code 2.18
Autocorrelation function
```
acf(resid)
```

2.5 Diagnostic Tests for Time Series Models

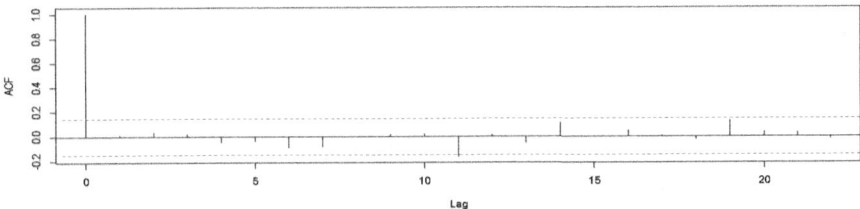

Fig. 2.20 ACF plot for the residuals of SARIMA(0,1,2)(0,1,3)$_{12}$ model

2.5.2 Diagnostic Tests for Homoscedasticity

The second assumption of a linear time series model is a constant variance (homoscedasticity) between the model residuals. Two diagnostic tests are used to diagnose homoscedasticity between the residuals. First, the ARCH-LM (autoregressive conditional heteroscedasticity-Lagrange multiplier) test proposed by Engle (1982) is used as a statistical method to determine whether the model residuals have a constant variance over time based on the F statistic. Second, the ACF plot and time plot of squared residuals are used as graphical methods to visually check for homoscedasticity.

2.5.2.1 ARCH-LM (Autoregressive Conditional Heteroscedasticity-Lagrange Multiplier) Test

Engle (1982) suggested the ARCH-LM test for investigating if the model residuals have a constant variance. The ARCH-LM test uses the F statistic for testing H_0: $a_i = 0$ ($i = 1,\ldots,m$) in the following linear regression of the squared model residuals:

$$\varepsilon_t^2 = a_0 + a_0 \varepsilon_{t-1}^2 + \ldots + a_0 \varepsilon_{t-m}^2 + e_t \qquad (2.21)$$

where ε_t^2 denotes the squared residuals at time t, e_t denotes the error term, and m represents the lag length of the squared residuals.

The null hypothesis of the ARCH-LM test is that the model residuals have constant variance (homoscedasticity); in other words, H_0: $a_0 = a_1 = \ldots = a_m = 0$ in Eq. 2.21. Rejection of the null hypothesis indicates the squared residuals have a significant relationship with each other, and a variable variance (heteroscedasticity) exists over time.

R Code 2.19 is used to test the homoscedasticity of the residuals of the SARIMA(0,1,2)(0,1,3)$_{12}$ using the ARCH-LM test. The homoscedasticity of the residuals was checked at different lag lengths (6, 12, 18, 24, 30, and 36 lag lengths).

R Code 2.19
ARCH test

```
library(MTS)
archTest(resid, lags=6)
archTest(resid, lags=12)
archTest(resid, lags=18)
archTest(resid, lags=24)
archTest(resid, lags=30)
archTest(resid, lags=36)
```

Table 2.11 The ARCH-LM test results for SARIMA(0,1,2)(0,1,3)$_{12}$ model residuals

Order	Test statistics	p-value
12	23.68	0.02
24	59.44	0.00
36	86.21	0.00

Table 2.11 shows the results of the ARCH-LM tests for the residuals of SARIMA(0,1,2)(0,1,3)$_{12}$ at different lag lengths. The multiple numbers of 12 are selected for the lag length because the monthly time series of HCS has a seasonal pattern for 12 months. The results show that the residuals are heteroscedastic at 12, 24, and 36 lag lengths at the 5% significance level.

2.5.2.2 ACF and Time Plots of Squared Residuals

Autocorrelation function and time plots of squared residuals are visually helpful for checking the homoscedasticity of a time series. R Code 2.20 is used to plot the ACF and time plots for the squared residuals of the SARIMA(0,1,2)(0,1,3)$_{12}$.

R Code 2.20
ACF and time plots of squared residuals

```
acf2(resid^2, max.lag=36)    #ACF plot of squared residuals
plot(resid^2, ylab="Squared Residuals")   #Time plot of squared residuals
```

2.5 Diagnostic Tests for Time Series Models

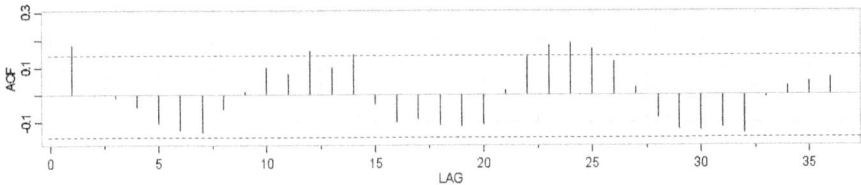

Fig. 2.21 ACF plot of squared residuals of the SARIMA(0,1,2)(0,1,3)$_{12}$ model

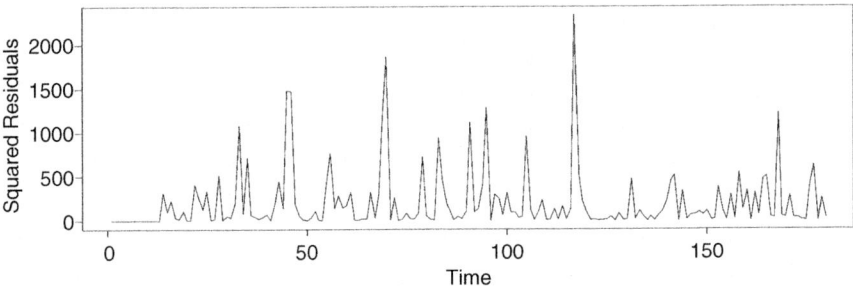

Fig. 2.22 Time plot of the squared residuals of the SARIMA(0,1,2)(0,1,3)$_{12}$ model

Figure 2.21 illustrates the ACF plot for the squared residuals of the SARIMA(0,1,2)(0,1,3)$_{12}$ model. The ACF plot of the squared residuals exhibits significant autocorrelations between the squared residuals at several lags, such as at lag 1. Also, the seasonal pattern is observed repetitively per 12 lag lengths. The significant relationship between the squared residuals in Fig. 2.21 indicates an inconstant variance (heteroscedasticity) in the time series over time.

Figure 2.22 illustrates a time plot of the squared residuals of the SARIMA(0,1,2)(0,1,3)$_{12}$ model. The time plot of squared residuals exhibits volatility clustering that signifies an inconstant variance (heteroscedasticity) through time.

2.5.3 Diagnostic Tests for Normality

Two diagnostic tests are introduced to evaluate the normality of model residuals. First, the Shapiro-Wilk test is used as a statistical method to check whether the residuals follow a normal distribution. Second, a normal quantile-quantile (Q-Q) plot is used as a graphical approach to diagnose the normal distribution of the residuals.

2.5.3.1 Shapiro-Wilk Test

Shapiro-Wilk test is widely used to diagnose the normality of the model residuals. The null hypothesis of the Shapiro-Wilk test is that the residuals follow a normal distribution. The rejection of the null hypothesis indicates that the residuals do not follow a normal distribution. R Code 2.21 is used to test the normality of the residuals of the SARIMA$(0,1,2)(0,1,3)_{12}$ using the Shapiro-Wilk test.

R Code 2.21
Shapiro-Wilk test

```
shapiro.test(resid)
```

Table 2.12 The results of Shapiro-Wilk tests for SARIMA$(0,1,2)(0,1,3)_{12}$ model residuals

W-statistic	p-value
0.9818	0.02

Table 2.12 shows the results of the Shapiro-Wilk test for the SARIMA$(0,1,2)(0,1,3)_{12}$ model. The results show that the null hypothesis is rejected. Hence, the residuals of the SARIMA$(0,1,2)(0,1,3)_{12}$ model do not follow a normal distribution.

2.5.3.2 Normal Q-Q (Quantile-Quantile) Plot

A normal Q-Q (quantile-quantile) plot is a visual method to assess if a set of data follow the normal distribution. The normal Q-Q plot illustrates the distribution of data against the normal distributions. Observations should lie approximately on a straight line if the data came from a normal distribution; otherwise, the data is not normally distributed. Model residuals follow a normal distribution if the quantile points reside along the reference line. Otherwise, the residuals are not normally distributed. R Code 2.22 is used to examine the normality of the residuals of the SARIMA$(0,1,2)(0,1,3)_{12}$ using a normal Q-Q plot.

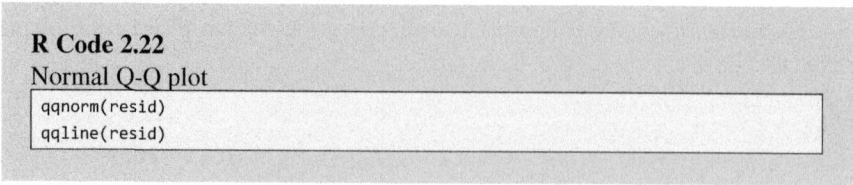

R Code 2.22
Normal Q-Q plot

```
qqnorm(resid)
qqline(resid)
```

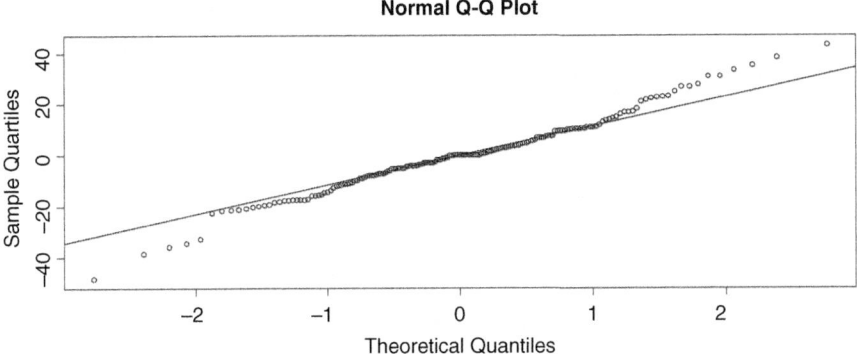

Fig. 2.23 Normal Q-Q plot of the SARIMA(0,1,2)(0,1,3)$_{12}$ model residuals

Figure 2.23 presents a normal Q-Q plot illustrating the distribution of the SARIMA(0,1,2)(0,1,3)$_{12}$ model residuals against the normal distribution. The quantile points are not clustered on the 45-degree reference line, especially at both ends of the line in Fig. 2.23. It suggests that the residuals of the SARIMA(0,1,2)(0,1,3)$_{12}$ are not normally distributed.

2.6 Summary

This chapter presented univariate time series forecasting models to predict construction time series. It was shown how to investigate the attributes of a time series (i.e., stationarity, trends, seasonality, etc.) and preprocess time series data (i.e., differencing) for forecasting. Several univariate time series forecasting models (i.e., moving average, autoregressive, exponential smoothing, autoregressive integrated moving average, and seasonal autoregressive integrated moving average models) are explained and compared. Diagnostic tests are discussed to investigate the existence of autocorrelation and heteroscedasticity in the residuals of the models and test their normality.

2.7 Problems

1. Apply ACF and PACF on the IHC time series and determine the AR and MA orders.
2. Apply the augmented Dickey-Fuller (ADF) test to the second-order differenced NHCCI time series, and report the results.
3. The ACF and PACF plots of a time series are provided below:
   (a) Determine whether the time series follows MA or AR process.
   (b) Determine the order of the process.

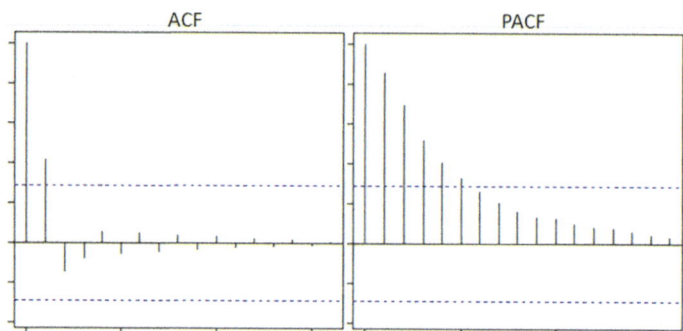

4. Develop an MA model to forecast CCCI for 12 months (2018), and compare the results with the existing AR model.
5. Test the normality of the model residuals for the trained MA model in Problem 4 (Chapter 2).
6. Develop single, double, and triple exponential smoothing models to forecast HCS for 12 months (2018), and compare the results with the SARIMA model.
7. Conduct the ARCH test to investigate the heteroscedasticity in the models trained in Problem 6 of Chapter 2.

References

Box, G. E., & Jenkins, G. M. (1976). *Time series analysis: Forecasting and control*. Holden-Day.
Brockwell, P. J., & Davis, R. A. (2016). *Introduction to time series and forecasting*. Springer.
California Department of General Services. (2019). *DGS California Construction Cost Index CCCI*, available at: https://www.dgs.ca.gov/RESD/Resources/Page-Content/Real-Estate-Services-Division-Resources-List-Folder/DGS-California-Construction-Cost-Index-CCCI. Classical and modern. Univ. of the West Indies Press.
Engle, R. F. (1982). Autoregressive conditional heteroscedasticity with estimates of the variance of United Kingdom inflation. Econometrica: *Journal of the econometric society*, 987–1007.
Federal Highway Administration (FHWA). (2019). *National Highway Construction Cost Index (NHCCI)*, available at: https://www.fhwa.dot.gov/policy/otps/nhcci/desc.cfm
Harvey, A. C. (1985). Trends and cycles in macroeconomic time series. *Journal of Business & Economic Statistics, 3*(3), 216–227.
Heckert, N. A., Filliben, J. J., Croarkin, C. M., Hembree, B., Guthrie, W. F., Tobias, P., & Prinz, J. (2002). *Handbook 151: NIST/SEMATECH e-Handbook of statistical methods*. National Institute of Standards and Technology.
Ho, S. L., Xie, M., & Goh, T. N. (2002). A comparative study of neural network and Box-Jenkins ARIMA modeling in time series prediction. *Computers & Industrial Engineering, 42*(2–4), 371–375.
Hyndman, R. J., & Athanasopoulos, G. (2018). *Forecasting: Principles and practice*. OTexts.
IOWA Department of Transportation (IOWADOT) (2019). *Price Trend Index for Iowa Highway Construction*, retrieve on December 2020, available at: https://www.iowadot.gov/contracts/lettings/PriceTrendIndex.pdf
Kim, S., Abediniangerabi, B., & Shahandashti, M. (2020). Forecasting pipeline construction costs using time series methods. In *Pipelines 2020* (pp. 198–209). American Society of Civil Engineers.

Nau, R. (2017). ARIMA models for time series forecasting. In *Statistical forecasting: Notes on regression and time series analysis*. Duke University. https://people.duke.edu/~rnau/Notes_on_nonseasonal_ARIMA_models%2D%2DRobert_Nau.pdf

Palachy, S. (2019). *Stationarity in time series analysis*. Towards Data Science. Saatavissa: https://towardsdatascience.com/stationarity-in-time-seriesanalysis-90c94f27322. Hakupäivä, 31, 2019.

Schmidt, A. F., & Finan, C. (2018). Linear regression and the normality assumption. *Journal of Clinical Epidemiology, 98*, 146–151.

Schwert, G. W. (2002). Tests for unit roots: A Monte Carlo investigation. *Journal of Business & Economic Statistics, 20*(1), 5–17.

Tukey, J. W. (1961). Discussion, emphasizing the connection between analysis of variance and spectrum analysis. *Technometrics, 3*(2), 191–219.

U.S. Census Bureau. (2019). *Total construction spending: Highway and street*, available at: http://www.census.gov/construction/c30/c30index.html

Watson, P. K., & Teelucksingh, S. S. (2002). *A practical introduction to econometric methods: Classical and modern*. University of West Indies Press.

Chapter 3
Construction Forecasting Using Time Series Volatility Models

Abstract Most linear time series models, including the univariate time series models, assume a time series has a constant variance over time. However, many construction time series data have not shown a constant variance. The volatility of a construction time series variable over time is challenging for accurate forecasting and risk management. This chapter discusses two time series volatility models (i.e., ARCH and GARCH) to forecast the variance of a construction time series. The ARCH and GARCH models are developed for modeling the volatility of the total federal construction spending time series published by the US Census Bureau. Then, the ARCH and GARCH models are combined with the ARIMA model to jointly estimate the mean and error variance of the total federal construction spending time series. The results show that the ARIMA-ARCH and ARIMA-GARCH models assuming a time-varying variance outperform the ARIMA model assuming a constant variance in terms of accuracy. R code examples are provided to develop time series volatility models and forecast a time series considering its time-varying variance. Exercise problems are presented at the end of the chapter for readers to review and practice the time series volatility models.

Keywords Time series volatility model · Construction forecasting · Homoscedasticity · Heteroscedasticity · ARCH (autoregressive conditional heteroscedasticity) model · GARCH (general ARCH) model · ARIMA-ARCH model · ARIMA-GARCH model · ARCH effects · Conditional variance · Weighted Ljung-Box test · LM-ARCH test · Time-varying variance

3.1 Introduction

Simple linear time series models assume that the model residuals are white noise. White noise residuals indicate that the statistical properties of residuals are constant over time. However, Engle (1982) found that the squared values of the residuals from the time series model to estimate the UK inflation show strong signs of autocorrelation, while the mean of the residuals is equal to 0. This finding indicates that the variance of the model residuals is not constant over time. The time-varying variance of the residuals violates the assumption of white noise residuals from simple linear time series models. Thus, time series volatility models are proposed to estimate and forecast the time-varying variance.

This chapter explains time series volatility models, including autoregressive conditional heteroscedasticity (ARCH) and general ARCH (GARCH), which are prevalently used to forecast the time-varying variance. In this chapter, seasonally adjusted values of total federal construction (TFC) are used to develop time series volatility models. The value of the total federal construction time series is a monthly value of federal construction put in place from 1993 to 2019 in the United States, published by the US Census Bureau (U.S. Census Bureau, 2022). The TFC time series measures monthly monetary values of total federal construction in millions of dollars and explains federal construction market conditions over time.

In the first step to developing a time series volatility model, a linear time series model is approximated to the TFC time series to capture the conditional mean of the time series over time. Then, the residuals of the linear time series model are examined to diagnose whether the residuals have a time-varying variance using the heteroscedasticity test. If heteroscedasticity (time-varying variance) exists, the time series volatility models are fitted to model the movements of the time-varying variance in the residuals of the linear time series model. After that, the diagnostic tests are conducted to check whether the assumptions of time series volatility models are satisfied. For example, the heteroscedasticity test is performed to check whether the residuals of the time series volatility models do not result in a time-varying variance.

3.2 Time Series Volatility Models

Time series volatility models are developed to approximate a time series whose variance changes over time. The following sections introduce two popular parametric time series volatility models: autoregressive conditional heteroscedasticity (ARCH) and generalized autoregressive conditional heteroscedasticity (GARCH) models.

3.2.1 Autoregressive Conditional Heteroscedasticity (ARCH)

In case a time series model does not satisfy an assumption of a constant variance over time, Engle (1982) suggested the autoregressive conditional heteroscedasticity (ARCH) model to simultaneously forecast the conditional mean and the error

3.2 Time Series Volatility Models

variance of a time series. The ARCH model applies an autoregressive process to forecast the time-varying variance instead of assuming a constant variance.

A forecast from an AR(1) model (i.e., $y_t = a_0 + a_1 y_{t-1} + \varepsilon_t$) has the conditional mean of $E_t(y_t) = a_0 + a_1 y_{t-1}$ and the error variance of

$$E_t\left[(y_t - a_0 - a_1 y_{t-1})^2\right] = E_t\left(\varepsilon_t^2\right) = \hat{\varepsilon}_t^2.$$

The ARCH model assumes that the error variance ($\hat{\varepsilon}_t^2$) at time t follows an AR(q) process using the past observations of the squared residuals. The ARCH(q) model to forecast the error variance at t is followed by:

$$E_t\left(\varepsilon_t^2\right) = \hat{\varepsilon}_t^2 = a_0 + a_1 \hat{\varepsilon}_{t-1}^2 + a_2 \hat{\varepsilon}_{t-2}^2 + \cdots + a_q \hat{\varepsilon}_{t-q}^2 \tag{3.1}$$

In the first step of developing an ARCH model, a linear ARIMA model is used to estimate the mean of the time series. Then, the residuals of the ARIMA model are used to test for heteroscedasticity. If heteroscedasticity is found in the model residuals, an ARCH model is developed to estimate the error variance.

R Code 3.1 can be used to develop an ARIMA model to specify a mean equation for the TFC time series. The training dataset of TFC is the monthly time series from January 1993 to December 2018. The testing dataset of TFC is 12 monthly time series from January 2019 to December 2019. Table 3.1 presents the result of the augmented Dickey-Fuller (ADF) test. The null hypothesis of the ADF test for TFC is not rejected at the 5% significance level. The result of the ADF test indicates that the training dataset of TFC is not stationary and requires first-order differencing to become a stationary time series.

R Code 3.1

ARIMA model for the mean of the TFC time series

```
[1] Forecasting performances of ARIMA-ARCH(12)
                ME     RMSE     MAE      MPE      MAPE     ACF1      Theil's U
Test set  -1400.374 1660.458 1400.374 -5.869887 5.869887 -0.3468954  14.59776

[2] Forecasting performances of ARIMA-GARCH(1,1)
                ME     RMSE     MAE      MPE      MAPE     ACF1      Theil's U
Test set  -1293.084 1575.462 1293.084 -5.397525 5.397525 -0.3286083  14.54813

[3] Forecasting performances of ARIMA(3,1,1)
                ME     RMSE     MAE      MPE      MAPE     ACF1      Theil's U
Test set  -1501.775 1744.319 1501.775 -6.32028  6.32028  -0.3661049  16.81901
```

Table 3.1 The result of the ADF test for the training dataset of total federal construction (TFC) value

Time series	Dickey-Fuller	Lag order	p-value
TFC	−1.22	6	0.90
ΔTFC	−6.85	6	0.01

Figure 3.1 shows the ACF and PACF plots for the first-differenced TFC time series. The ACF plot shows a spike at lag 1, suggesting the MA(q) order of 1. The PACF plot presents a spike at lag 3, indicating the AR(p) order to be 3. The ARIMA(3,1,1) is selected to model the TFC time series based on the observations of ACF and PACF plots.

Figure 3.2 illustrates the actual observations, the fitted values, and the model residuals of the ARIMA(3,1,1) model for the TFC time series.

The ARCH test can be used to test for the presence of conditional heteroscedasticity (i.e., ARCH effects) in the residuals of the ARIMA(3,1,1) model. The ARCH test examines the parameters in the AR(q) process of the squared residuals ($\hat{\varepsilon}_t^2$) as follows:

$$\hat{\varepsilon}_t^2 = a_0 + a_1\hat{\varepsilon}_{t-1}^2 + a_2\hat{\varepsilon}_{t-2}^2 + \cdots + a_q\hat{\varepsilon}_{t-q}^2 \tag{3.2}$$

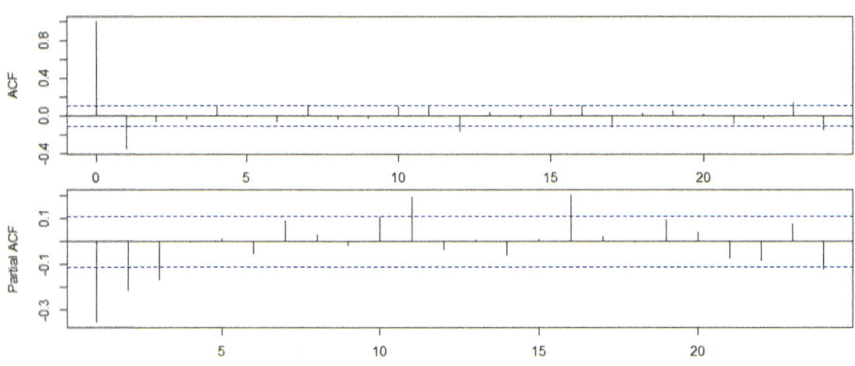

Fig. 3.1 The ACF and PACF plots for the TFC time series

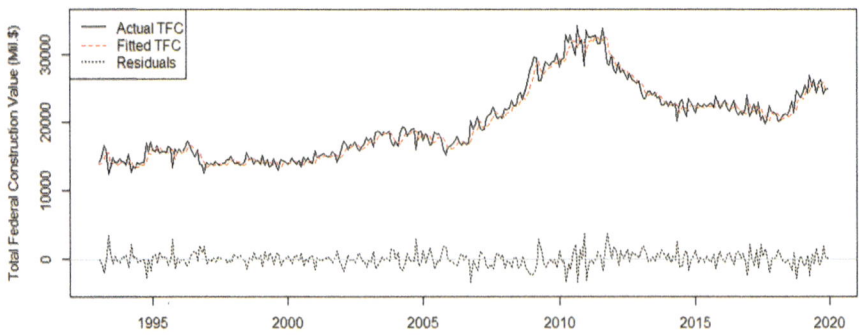

Fig. 3.2 Actual, fitted, and residuals of the ARIMA(3,1,1) model for the TFC time series

3.2 Time Series Volatility Models

The null hypothesis of the ARCH test is $a_1 = a_2 \cdots = a_q = 0$. R Code 3.2 can be used to conduct the ARCH test and plot the ACF and PACF for the residuals of the ARIMA(3,1,1) model to check for heteroscedasticity.

R Code 3.2

ARCH test and PACF plot to check for heteroscedasticity

```
library(ConstructionAnalyticsR)
tfc <- ConstructionAnalyticsR::tfc
tfc.ts <- ts(tfc, frequency = 12, start = c(1993,1), end = c(2019,12))
tfc_train <- ts(tfc[1:312], frequency = 12, start = c(1993,1), end =c(2018,12))
tfc_test <- ts(tfc[313:324], frequency = 12, start = c(2019,1), end =c(2019,12))
plot.ts(tfc.ts, main = "Time-series of Total Federal Construction Value", xlab = "Years",
ylab = "TFC")

library(astsa)
adf.test(tfc_train)
dtfc_train <- diff(tfc_train)
adf.test(dtfc_train)
acf2(dtfc_train)
tfc_arima=arima(tfc_train,order=c(3,1,1))
```

Table 3.2 The result of the ARCH test for the residuals of ARIMA(3,1,1)

Chi-squared statistic	Degree of freedom	p-value
27.79	12	0.01

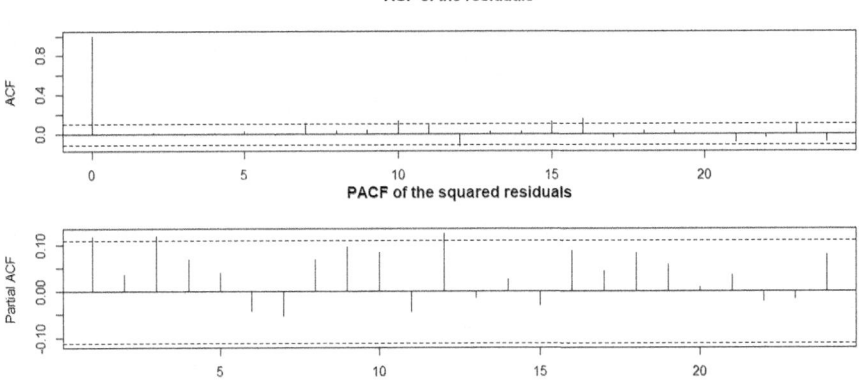

Fig. 3.3 The ACF for the residuals and PACF for the squared residuals of the ARIMA(3,1,1) model

Table 3.2 presents the result of the ARCH test for the ARIMA(3,1,1) model residuals. The result shows the rejection of the null hypothesis at the 1% significance level, indicating heteroscedasticity in the ARIMA(3,1,1) model residuals.

The ACF for the residuals and PACF for the squared residuals of the ARIMA(3,1,1) model are plotted in Fig. 3.3. The ACF plot of the residuals shows that the residuals have no serial correlation. The PACF plot of the squared residuals exhibits considerable spikes, indicating significant partial autocorrelations between

R Code 3.3

ARCH model selection based on AICc

```
library(FinTS)
ArchTest(tfc_arima$residuals) # ARCH test for the residuals of ARIMA(3,1,1)
acf(tfc_arima$residuals) # ACF for the residuals of ARIMA(3,1,1)
pacf(tfc_arima$residuals^2) # PACF for the squared residuals of ARIMA(3,1,1)
```

Table 3.3 The log-likelihood and AICc for ARCH models

ARCH model	Log-likelihood	AICc
ARCH(1)	−2615.772	5235.584
ARCH(3)	−2594.942	5198.013
ARCH(12)	−2507.548	5042.318

the squared model residuals at lag 1, 3, and 12. The time-varying variance (heteroscedasticity) was found in the residuals of the ARIMA(3,1,1) model for TFC. Thus, the ARCH time series volatility model is applied to estimate and forecast the variance of the model residuals.

The PACF plot in Fig. 3.3 shows spikes at lag 1, 3, and 12. The result of the ARCH-LM test identifies the conditional heteroscedasticity among the ARIMA(3,1,1) model residuals in Table 3.2.

R Code 3.4

ARCH(12) model

```
# Creating ARCH models
library(tseries)
arch1 <- garch(tfc_arima$residuals, c(0,1), trace=FALSE)
arch3 <- garch(tfc_arima$residuals, c(0,3), trace=FALSE)
arch12 <- garch(tfc_arima$residuals, c(0,12), trace=FALSE)
N <- length(tfc_arima$residuals)
loglik1 <- logLik(arch1)
loglik3 <- logLik(arch3)
loglik12 <- logLik(arch12)
loglik <- c(loglik1, loglik3, loglik12)
q <- c(1, 3, 12)
k <- q + 1
aicc <- -2 * loglik + 2 * k * N / (N - k - 1)
print(data.frame(q, loglik, aicc)) # Selecting ARCH models based on AICc
```

It is generally recommended to choose the ARCH model based on the PACF plot and the information criteria, such as the Akaike information criterion with correction (AICc) (Brockwell & Davis, 2016; Tsay, 2010). Since the PACF plot in Fig. 3.3 shows spikes at lag 1, 3, and 12, the AICc values of ARCH(1), ARCH(3), and ARCH(12) models are compared using R Code 3.3. The ARCH model with the lowest AICc value will be selected.

3.2 Time Series Volatility Models

Table 3.4 The optimal parameters for the ARCH(12) model

	Estimate	Standard error	t-value	p-value
alpha1	0.19	0.08	2.3	0.02
alpha2	0.06	0.04	1.48	0.14
alpha3	0.12	0.06	2.06	0.04
alpha4	0.04	0.05	0.92	0.36
alpha5	0.00	0.05	0.01	0.99
alpha6	0.07	0.06	1.28	0.20
alpha7	0.11	0.03	3.43	0.00
alpha8	0.07	0.05	1.42	0.16
alpha9	0.04	0.01	3.67	0.00
alpha10	0.09	0.05	2.05	0.04
alpha11	0.00	0.17	0.00	1.00
alpha12	0.21	0.04	4.78	0.00

Table 3.3 shows the log-likelihood and AICc values for ARCH(1), ARCH(3), and ARCH(12). Since the ARCH(12) model provides the lowest value of AICc and the highest value of log-likelihood, the ARCH(12) model is selected to fit the variance of the ARIMA(3,1,1) model residuals.

The ARCH(12) model for the error variance of the ARIMA(3,1,1) can be developed using R Code 3.4. The residuals of the ARIMA(3,1,1) were found to approximately follow a skewed student-t distribution.

The ARIMA(3,1,1) model for the conditional mean of TFC and the ARCH(12) model for the residuals of ARIMA(3,1,1) are represented by:

$$\Delta s_t = -0.49\Delta s_{t-1} - 0.3\Delta s_{t-2} - 0.17\Delta s_{t-3} + 0.02\varepsilon_{t-1} + \varepsilon_t \tag{3.3}$$

$$\varepsilon_t^2 = 0.19\varepsilon_{t-1}^2 + 0.06\varepsilon_{t-2}^2 + 0.12\varepsilon_{t-3}^2 + 0.04\varepsilon_{t-4}^2 + 0.07\varepsilon_{t-6}^2 \\ + 0.11\varepsilon_{t-7}^2 + 0.07\varepsilon_{t-8}^2 + 0.04\varepsilon_{t-9}^2 + 0.09\varepsilon_{t-10}^2 + 0.21\varepsilon_{t-12}^2 \tag{3.4}$$

where Δs_t is the first-differenced TFC series forecast at time t and ε_t^2 is the error variance forecast at time t.

Table 3.4 presents the optimal parameters for the ARCH(12) model based on the maximum Gaussian likelihood function.

3.2.2 Generalized Autoregressive Conditional Heteroscedasticity (GARCH)

The ARCH model helps forecast the error variance in various time series. However, the ARCH model often needs to approximate the structure of a long lag length to capture the information in the variance. The AR process with a long lag length of the squared residuals can harm the parsimony of the model, requiring many parameters to be estimated. Bollerslev (1986) proposed a parsimonious generalized

ARCH(GARCH) model by developing a technique that allows the variance to follow an ARMA process. The generalized autoregressive conditional heteroscedasticity (GARCH) model includes both autoregressive (AR) and moving average (MA) components for the time-varying variance of error terms. The error process comprises two terms, as expressed by Eq. (3.5).

$$\varepsilon_t = v_t \sqrt{h_t} \qquad (3.5)$$

where ε_t is the error term of the time series model, h_t is a variance of ε_t, and v_t is a white noise process with zero mean and unit variance ($\sigma_v^2 = 1$).

A GARCH(p,q) model is expressed by Eq. (3.6).

$$h_t = a_0 + \sum_q^{j=1} a_j \varepsilon_{t-j}^2 + \sum_p^{i=1} b_i h_{t-i} \qquad (3.6)$$

where h_t denotes a conditional variance, a_0 is the intercept, ε_{t-j}^2 ($j = 1,\ldots,q$) is the squared residuals of the time series model at time $t - j$, p is the AR order of the conditional variance, and q is the MA order for the squared residuals.

A GARCH(p,q) model is a generalized ARCH(q) model. The ARCH(q) model can be expressed by GARCH(0,q) model. A parsimonious GARCH(1,1) model is successfully applied to approximate a higher-order ARCH process of the error variance in a time series (Miah & Rahman, 2016; Jafari et al., 2007; Bollerslev, 1986). Therefore, a GARCH(1,1) model is developed for TFC using R Code 3.5.
The developed GARCH(1,1) model using R Code 3.5. is represented by:

$$h_t = 0.05\varepsilon_{t-1}^2 + 0.95h_{t-1} \qquad (3.7)$$

where h_t is the variance at time t and ε_{t-1} is the residual of the ARIMA(3,1,1) model for TFC.

Table 3.5 presents the optimal parameters for the GARCH(1,1) model based on the maximum Gaussian likelihood function. The parameters of the GARCH(1,1) model are statistically significant.

R Code 3.5

GARCH(1,1) model

```
# Calling library
library(rugarch)

# Creating ARCH(12) model
arch = ugarchspec()
arch=ugarchspec(variance.model = list(model="sGARCH", garchOrder=c(12,0)),
mean.model=list(include.mean=F,armaOrder=c(0,0)), distribution.model="sstd",
fixed.pars=list(omega=0))

# Fitting ARCH(12) model to the residuals of ARIMA(3,1,1)
ARCH12= ugarchfit(spec = arch, data = tfc_arima$residuals, solver='hybrid')
```

Table 3.5 The optimal parameters for the GARCH(1,1) model

	Estimate	Standard error	t-value	p-value
alpha1	0.05	0.02	2.77	0.01
beta1	0.95	0.02	48.20	0.00

3.3 Diagnostic Tests for Time Series Volatility Models

Time series volatility models are expected to estimate and forecast the dynamic movements of a time-varying variance in the residuals of simple linear time series models. Two diagnostic tests are conducted to check whether the residuals of time series volatility models satisfy the following assumptions for the time series volatility models:

- The residuals are not correlated. The serial correlation between residuals indicates that the information that can improve forecasting accuracy remains in the residuals.
- The residuals have a constant variance. If the residuals show a time-varying variance, the time series volatility models are not adequately developed to capture the ARCH effects between residuals.

When the time series volatility models satisfy these assumptions, they are adequately implemented to forecast the error variance. First, the Ljung-Box test is used to check for no autocorrelation between the model residuals. Second, the autoregressive conditional heteroscedasticity (ARCH) test is conducted to examine the assumption of a constant variance (homoscedasticity). The standardized model residuals of time series volatility models developed in the previous sections are diagnosed using the diagnostic tests in the following sections.

3.3.1 Results of Diagnostic Tests for No Autocorrelation

Fisher and Gallagher (2012) found that the weighted portmanteau statistics provide more stable results for the diagnostic test on the standardized residuals to check the adequacy of a fitted time series volatility model. R Code 3.6 can be used to conduct the weighted Ljung-Box tests to diagnose no autocorrelation between the standardized residuals of the developed time series volatility models.

Table 3.6 presents the results of the Ljung-Box tests on the standardized residuals of the ARCH(12) and GARCH(1,1) models. Since the null hypothesis of no autocorrelation is not rejected at the 5% significance level, all the developed time series volatility models satisfy the assumption of no autocorrelation between the model residuals.

R Code 3.6

Diagnostic tests for no autocorrelation

```
# Creating GARCH(1,1) model
garch = ugarchspec()
garch=ugarchspec(variance.model = list(model="sGARCH", garchOrder=c(1,1)),
mean.model=list(include.mean=F,armaOrder=c(0,0)), distribution.model="sstd",
fixed.pars=list(omega=0))

# Fitting GARCH(1,1) model to the residuals of ARIMA(3,1,1)
GARCH11= ugarchfit(spec = garch, data = tfc_arima$residuals, solver='hybrid')
```

Table 3.6 The results of the Ljung-Box test on the standardized residuals of the developed models

Model	Weighted X-squared	p-value
ARCH(12)	0.15	0.69
GARCH(1,1)	0.02	0.89

3.3.2 Results of Diagnostic Tests for Homoscedasticity

The LM-ARCH tests are conducted using R Code 3.7 to diagnose the assumption of homoscedasticity for the developed volatility models.

Table 3.7 shows the results of the ARCH tests for the ARCH(12) and GARCH(1,1) volatility models. The null hypothesis of homoscedasticity is not rejected at the 5% significance level for the developed volatility models. According to the results of the ARCH tests, there exists no heteroscedasticity between the residuals of the developed volatility models.

R Code 3.7

Diagnostic tests for homoscedasticity

```
# Standardized residuals
ARCHstdres= ARCH12@fit$residuals/(ARCH12@fit$var^(1/2))
GARCHstdres= GARCH11@fit$residuals/(GARCH11@fit$var^(1/2))

# Weighted Ljung-Box tests
library(WeightedPortTest)
Weighted.Box.test(ARCHstdres, type="Ljung-Box")
Weighted.Box.test(GARCHstdres,type="Ljung-Box")
```

Table 3.7 The results of the ARCH-LM tests for the developed time series volatility models

	Weighted chi-squared	p-value
ARCH(12)	3.51	0.84
GARCH(1,1)	2.57	0.94

3.4 Estimating Volatility Using ARCH and GARCH Models

This section validates the performance of time series volatility models by comparing the observed and estimated volatilities using ARCH(12) and GARCH(1,1) models. The actual volatility is a latent variable and can be neither measured nor observed directly (Engle & Patton, 2007). Therefore, realized volatility is alternatively used to roughly estimate actual volatility and evaluate the performance of time series volatility models (Ilbeigi et al., 2017; Danielsson, 2011; Andersen & Bollerslev, 1998). The realized volatility is commonly estimated by the squares of the changes (Ilbeigi et al., 2017; Bollen & Whaley, 2015). Thus, the realized volatility of TFC is calculated by the squares of the TFC changes. R Code 3.8 is used to illustrate the comparison between the realized volatility measured by the squares of monthly changes in TFC and the estimated volatility using ARCH(12) and GARCH(1,1) models. Figure 3.4 shows a volatility estimation using ARCH(12) and GARCH(1,1) models and compares the realized volatility and the estimated volatility using ARCH(12) and GARCH(1,1) models.

R Code 3.8

Volatility estimation and comparison

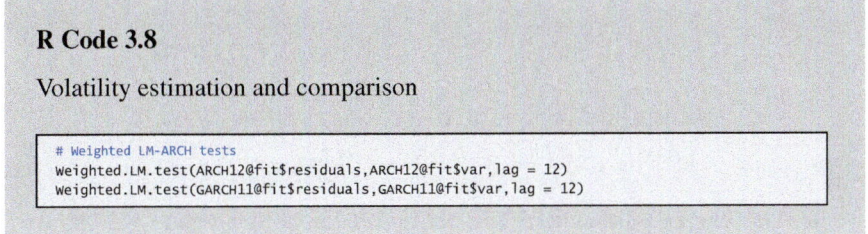

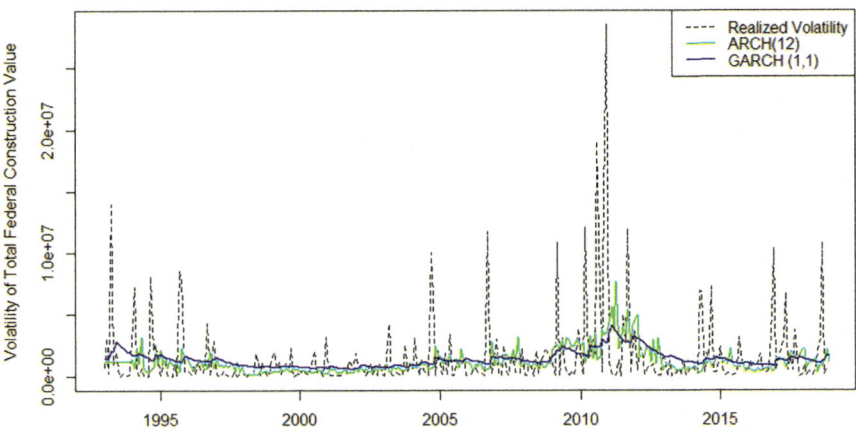

Fig. 3.4 Volatility estimation and comparison between the realized volatility and the estimated volatility using ARCH(12) and GARCH(1,1)

3.5 Forecasting the TFC Time Series Using ARIMA-ARCH and ARIMA-GARCH Models

Since time series volatility models, including ARCH and GARCH models, are fitted to the residuals of linear time series models such as ARIMA, time series volatility models can be combined with ARIMA models (Joukar & Nahmens, 2016). In this section, ARCH and GARCH models are combined with the ARIMA model to jointly estimate the mean and error variance of time series. Then, the forecasting performances of ARIMA-ARCH and ARIMA-GARCH models are compared with the performance of a single ARIMA model for forecasting 12-month-ahead TFC values in 2019. While the ARIMA(3,1,1) model assumes a constant variance in the model residuals, ARCH(12) and GARCH(1,1) models assume a time-varying variance in the model residuals and forecast the future variance using the relationship between the variances over time. The ARIMA(3,1,1)-ARCH(12) and ARIMA(3,1,1)-GARCH(1,1) models can be developed using R Code 3.9.

The developed ARIMA(3,1,1)-ARCH(12) model using R Code 3.9 is represented by:

$$\Delta s_t = 16.9 - 0.48\Delta s_{t-1} - 0.28\Delta s_{t-2} - 0.19\Delta s_{t-3} + \varepsilon_t + 0.01\varepsilon_{t-1} \tag{3.8}$$

$$\begin{aligned}\varepsilon_t^2 = &\ 0.19\varepsilon_{t-1}^2 + 0.06\varepsilon_{t-2}^2 + 0.12\varepsilon_{t-3}^2 + 0.05\varepsilon_{t-4}^2 + 0.08\varepsilon_{t-6}^2 + 0.1\varepsilon_{t-7}^2 \\ &+ 0.06\varepsilon_{t-8}^2 + 0.03\varepsilon_{t-9}^2 + 0.09\varepsilon_{t-10}^2 + 0.21\varepsilon_{t-12}^2\end{aligned} \tag{3.9}$$

where Δs_t is the first-differenced TFC series forecast at time t and ε_t^2 is the error variance forecast at time t.

R Code 3.9

ARIMA-ARCH(12) and ARIMA-GARCH(1,1) model development

```
realv=diff(tfc_train)^2 # Realized volatility
vARCH12=ts(ARCH12@fit$var,start=c(1993,1),frequency=12) # Estimated volatility using ARCH (12)
vGARCH11=ts(GARCH11@fit$var,start=c(1993,1),frequency=12) # Estimated volatility using GARCH (1,1)
plot(ts(realv,start=c(1993,1),frequency=12), main="Volatility of Total Federal Construction Value")
lines(vARCH12,col='green')
lines(vGARCH11,col='blue')
legend(legend=c("Realized Volatility","ARCH(12)","GARCH(1,1)"), lty=c(1,1,1),
col=c("black","green","blue"),"topright")
```

3.5 Forecasting the TFC Time Series Using ARIMA-ARCH and ARIMA-GARCH...

Table 3.8 The optimal parameters for the ARIMA(3,1,1)-ARCH(12) model

	Estimate	Standard error	t-value	p-value
mu	16.90	32.69	0.52	0.61
ar1	−0.48	0.24	−2.00	0.05
ar2	−0.28	0.11	−2.51	0.01
ar3	−0.19	0.07	−2.84	0.00
ma1	0.01	0.24	0.05	0.96
alpha1	0.19	0.11	1.66	0.10
alpha2	0.06	0.04	1.40	0.16
alpha3	0.12	0.06	1.92	0.06
alpha4	0.05	0.05	0.93	0.35
alpha5	0.00	0.11	0.00	1.00
alpha6	0.08	0.07	1.09	0.28
alpha7	0.10	0.07	1.45	0.15
alpha8	0.06	0.05	1.30	0.19
alpha9	0.03	0.03	1.02	0.31
alpha10	0.09	0.05	1.95	0.05
alpha11	0.00	0.10	0.00	1.00
alpha12	0.21	0.06	3.53	0.00

Table 3.8 presents the optimal parameters for the ARIMA(3,1,1)-ARCH(12) model based on the maximum Gaussian likelihood function.

The ARIMA(3,1,1)-GARCH(1,1) model developed using R Code 3.9 is represented by:

$$\Delta s_t = 26.69 - 0.47\Delta s_{t-1} - 0.3\Delta s_{t-2} - 0.16\Delta s_{t-3} + \varepsilon_t - 0.004\varepsilon_{t-1} \quad (3.10)$$

$$h_t = 0.05\varepsilon_{t-1}^2 + 0.95h_{t-1} \quad (3.11)$$

where Δs_t is the first-differenced TFC series forecast at time t and ε_t^2 is the error variance forecast at time t.

Table 3.9 presents the optimal parameters for the ARIMA(3,1,1)-GARCH(1,1) model based on the maximum Gaussian likelihood function. The parameters of the GARCH(1,1) model are statistically significant.

The ARIMA(3,1,1)-ARCH(12) and ARIMA(3,1,1)-GARCH(1,1) models forecast monthly series and variance for TFC in 2019 using R Code 3.10. The forecasts from ARIMA-ARCH and ARIMA-GARCH models are compared with the forecasts from the single ARIMA(3,1,1) model to evaluate the forecasting performances of ARIMA-ARCH and ARIMA-GARCH models.

Table 3.10 presents the actual values and forecasts for the TFC time series using the ARCH(12), GARCH(1,1), and ARIMA(3,1,1) models for 12 months in 2019.

Table 3.9 The optimal parameters for the GARCH(1,1) model

	Estimate	Standard error	t-value	p-value
mu	26.69	29.12	0.92	0.36
ar1	−0.47	0.29	−1.64	0.10
ar2	−0.30	0.14	−2.21	0.03
ar3	−0.16	0.08	−2.13	0.03
ma1	−0.004	0.29	−0.01	0.99
alpha1	0.05	0.02	2.73	0.01
beta1	0.95	0.02	48.33	0.00

R Code 3.10

Forecasting performances of ARIMA-ARCH and ARIMA-GARCH

```
# Developing ARIMA-ARCH(12)
m.arch = ugarchspec()

m.arch =ugarchspec(variance.model = list(model="sGARCH", garchOrder=c(12,0)),
mean.model=list(include.mean=T,armaOrder=c(3,1)), distribution.model="sstd",
fixed.pars=list(omega=0))

m.ARCH12= ugarchfit(spec = m.arch, data = dtfc_train, solver='hybrid')


# Developing ARIMA-GARCH(1,1)
m.garch = ugarchspec()

m.garch=ugarchspec(variance.model = list(model="sGARCH", garchOrder=c(1,1)),
mean.model=list(include.mean=T,armaOrder=c(3,1)), distribution.model="sstd",
fixed.pars=list(omega=0))

m.GARCH11= ugarchfit(spec = m.garch, data = dtfc_train, solver='hybrid')
```

Results

```
bootarch=ugarchboot(m.ARCH12,method=c("Partial","Full")[1],n.ahead=
12,n.bootpred=3000,n.bootfit=3000)  # Bootstrapping the ARIMA-ARCH
arch_s=bootarch@forc@forecast$seriesFor  # Forecasting monthly series using ARIMA-ARCH
arch_v=bootarch@forc@forecast$sigmaFor   # Forecasting monthly variances using ARIMA-ARCH
arch_ss<-ts(diffinv(as.vector(arch_s),difference=1,lag=1,xi=23602)[2:13],
start=c(2019,1),frequency = 12)

bootgarch=ugarchboot(m.GARCH11,method=c("Partial","Full")[1],n.ahead=12,n.bootpred=3000,n.
bootfit=3000)  # Bootstrapping the ARIMA-GARCH
garch_s=bootgarch@forc@forecast$seriesFor  # Forecasting monthly series using ARIMA-GARCH
garch_v=bootgarch@forc@forecast$sigmaFor   # Forecasting monthly variances using ARIMA-GARCH
garch_ss<-ts(diffinv(as.vector(garch_s),difference=1,lag=1,xi=23602)[2:13],
start=c(2019,1),frequency = 12)

#Call the library
library(forecast)
arima_ss=ts(forecast(tfc_arima, h=12)$mean, start=c(2019,1),frequency = 12)

#Calcutlate the prediction performance metrics
accuracy(tfc_test,arch_ss)
accuracy(tfc_test,garch_ss)
accuracy(tfc_test,arima_ss)
```

3.6 Summary

Table 3.10 Actual values and forecasts of TFC for 12 months using ARIMA(3,1,1)-ARCH(12), ARIMA(3,1,1)-GARCH(1,1), and ARIMA(3,1,1) models

Month	Actual values	ARIMA-ARCH forecasts	ARIMA-GARCH forecasts	ARIMA forecasts
Jan	24414	23493.18	23584.94	23504.54
Feb	25437	23863.67	23934.63	23843.89
Mar	24327	23893	23953.4	23842.06
Apr	26960	23827.05	23894.62	23759.22
May	25150	23814.72	23910.91	23741.09
Jun	26238	23866.82	23969.35	23774.83
Jul	24307	23890.46	23998.2	23778.18
Aug	26058	23899.57	24016.1	23769.69
Sep	26316	23911.79	24041.08	23766.96
Oct	24161	23931.9	24070.85	23770.23
Nov	24775	23950.04	24098.04	23770.92
Dec	24970	23966.31	24123.86	23770.09

Table 3.11 Forecasting errors of ARIMA(3,1,1)-ARCH(12), ARIMA(3,1,1)-GARCH(1,1), and ARIMA(3,1,1) models

Models	MAPE	MAE	RMSE
ARIMA-ARCH	5.86	1400.37	1660.45
ARIMA-GARCH	5.39	1293.08	1575.46
ARIMA	6.32	1501.77	1744.31

Table 3.11 presents the forecasting error measures of ARIMA(3,1,1)-ARCH(12), ARIMA(3,1,1)-GARCH(1,1), and ARIMA(3,1,1) models. ARIMA-ARCH and ARIMA-GARCH models provide more accurate forecasts than the single ARIMA model. Thus, ARCH and GARCH models can improve forecasting accuracy by estimating and predicting the time-varying variance in the residuals of the linear time series model.

3.6 Summary

Linear time series models assume that the model residuals have constant statistical properties over time. However, it is not surprising to observe a time-varying variance in the residuals of linear time series models that are approximated to the real-life construction time series. This chapter presented time series volatility models to forecast the time-varying variance in the ARIMA model residuals. First, two popular parametric time series volatility models, including ARCH and GARCH models, are explained. Then, the ARCH and GARCH models are developed and diagnosed to check whether the model assumptions are satisfied. After conducting two diagnostic tests, the ARCH and GARCH models are implemented for forecasting the total federal construction (TFC) time series. The time series volatility models estimate and forecast the time-varying variance and provide information about the volatility of the time series. This information can be used for implementing risk-management strategies in the volatile and uncertain construction market, such as determining the proper risk premium at the right time and preparing for market volatility. Finally, a hybrid time series model combining ARCH and GARCH models with the ARIMA model is discussed and developed to measure both the mean and variance of the TFC time series. The hybrid ARIMA-ARCH and ARIMA-GARCH models adequately approximated the time-varying variance and improved forecasting accuracy over the single ARIMA model. These hybrid models can assist construction engineers and managers in forecasting the construction time series more accurately and estimate the time-varying variance. The time series volatility models are beneficial for practitioners to quantify and forecast the volatility, identify the current state of risk, and determine the timely risk management plans in construction projects.

3.7 Exercise Problems

1. Let $y_0 = 7$ and the first five realizations of the $\{\varepsilon_t\}$ sequence be $(-2, -1, 0, 1, 2)$. Calculate the sample mean and variance of $\{y_t\}$ in the following model:

$$y_t = 0.5 y_{t-1} + \varepsilon_t + \varepsilon_{t-1}^2$$

2. Interpret the ACF and PACF plots of the residuals and squared residuals below. Explain if there exist ARCH effects.

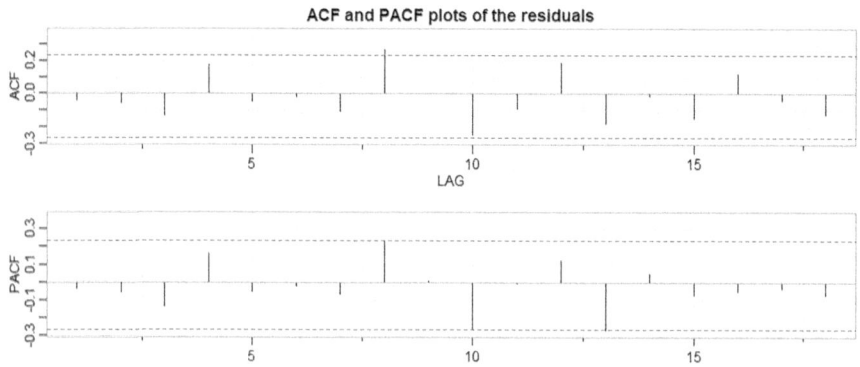

3.7 Exercise Problems

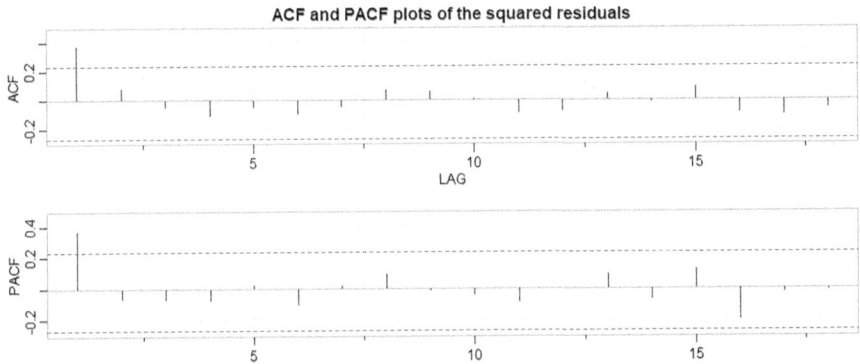

3. Use the HCS (Highway Construction Spending) time series ("hcs.csv" in the *ConstructionAnalyticsR* package). The rate of change in HCS (R_t) is approximately defined as the first log differences:

$$R_t = 100 \times \left(\log(HCS_t) - \log(HCS_{t-1}) \right)$$

(1) Plot graphs of the HCS time series, the rate of change in HCS, and the squared rate of change in HCS. Do the graphs show ARCH effects?

(2) Develop an ARMA model for R_t, and diagnose whether the developed model satisfies modeling assumptions.

(3) Test for ARCH effects in the residuals of the developed ARMA model in the previous question 3-(2). Discuss the results of the ARCH test.

(4) If heteroscedasticity is found in the model residuals, develop a time series volatility model (e.g., GARCH(1,1)) to estimate and forecast the time-varying variance.

(5) Plot the graphs of the residuals, the squared residuals, the conditional variance, and the conditional standard deviation. Compare the conditional standard deviation with the residuals, and discuss whether heteroscedasticity remains in the residuals based on the plot at the 95% confidence interval. The 95% confidence interval for the residuals is as follows:

$$\max = 1.96 \times \text{conditional standard deviation}$$

$$\min = -1.96 \times \text{conditional standard deviation}$$

(6) Develop a hybrid ARIMA-GARCH model for R_t. Forecast 12 out-of-sample values and error variances of HCS. Compare forecasting accuracy between the hybrid ARIMA-GARCH model and the single ARIMA model.

(7) Forecast 1000 out-of-sample values and error variances of HCS using the hybrid ARIMA-GARCH model in the previous question 3-(6). Also, forecast 10,000 out-of-sample values and error variances of HCS. Interpret and compare the results between the 12-period, 1000-period, and 10,000-period out-of-sample forecasting.

References

Andersen, T. G., & Bollerslev, T. (1998). Answering the skeptics: Yes, standard volatility models do provide accurate forecasts. *International Economic Review, 39*, 885–905.

Bollen, N. P., & Whaley, R. E. (2015). Futures market volatility: What has changed? *Journal of Futures Markets, 35*(5), 426–454.

Bollerslev, T. (1986). Generalized autoregressive conditional heteroskedasticity. *Journal of Econometrics, 31*(3), 307–327.

Brockwell, P. J., & Davis, R. A. (2016). *Introduction to time-series and forecasting*. Springer.

Danielsson, J. (2011). *Financial risk forecasting: The theory and practice of forecasting market risk with implementation in R and Matlab* (Vol. 588). Wiley.

Engle, R. F. (1982). Autoregressive conditional heteroscedasticity with estimates of the variance of United Kingdom inflation. *Econometrica, 50*(4), 987–1007.

Engle, R. F., & Patton, A. J. (2007). What good is a volatility model? In *Forecasting volatility in the financial markets* (pp. 47–63). Butterworth-Heinemann.

Fisher, T. J., & Gallagher, C. M. (2012). New weighted portmanteau statistics for time-series goodness of fit testing. *Journal of the American Statistical Association, 107*(498), 777–787.

Ilbeigi, M., Castro-Lacouture, D., & Joukar, A. (2017). Generalized autoregressive conditional heteroscedasticity model to quantify and forecast uncertainty in the price of asphalt cement. *Journal of Management in Engineering, 33*(5), 04017026.

Jafari, G. R., Bahraminasab, A., & Norouzzadeh, P. (2007). Why does the standard GARCH(1, 1) model work well? *International Journal of Modern Physics C, 18*(07), 1223–1230.

Joukar, A., & Nahmens, I. (2016). Volatility forecast of construction cost index using general autoregressive conditional heteroskedastic method. *Journal of Construction Engineering and Management, 142*(1), 04015051.

Miah, M., & Rahman, A. (2016). Modelling volatility of daily stock returns: Is GARCH(1,1) enough? *American Scientific Research Journal for Engineering, Technology, and Sciences (ASRJETS), 18*(1), 29–39.

Tsay, R. S. (2010). *Analysis of financial time-series*. Wiley.

U.S. Census Bureau, Construction Spending. (2022, September 1). *Additional information on the survey methodology*. Retrieved from www.census.gov/construction/c30/meth.html

Chapter 4
Construction Time Series Forecasting Using Multivariate Time Series Models

Abstract Identifying leading indicators of construction cost time series and using them as explanatory variables could improve the accuracy of forecasting models. This chapter explains the process of identifying the leading indicators of a construction time series and developing proper multivariate models, such as vector error correction and vector autoregressive models for forecasting them. Several practical examples are provided along with R codes to show how to create and diagnose multivariate time series models for forecasting construction variables. A multivariate time series model is developed for forecasting monthly Highway Construction Spending (HCS) time series using consumer price index (CPI) as the leading indicator, and its performance is compared with the results of the univariate seasonal ARIMA model. The comparison results show that the VEC model outperforms the seasonal autoregressive integrated moving average (SARIMA) model based on typical error measures.

Keywords Construction cost forecasting · Multivariate time series models · Explanatory time series · Granger causality · Cointegration test · Vector autoregressive · Vector error correction

4.1 Introduction

Multivariate time series models use autocorrelation and cross-correlation in multiple time series to improve the forecasting measures, such as in-sample and out-of-sample accuracies, compared to univariate time series modeling that only uses the historical information in one time series for forecasting. In this section, two multivariate time series models (i.e., vector error correction (VEC) and vector autoregressive (VAR)) are explained and used for forecasting construction time series. The following steps are described to create multivariate time series models:

- Examining the series' main characteristics, such as stationarity, trend, seasonality, sharp changes in behavior, and outlying observations

Table 4.1 HCS potential explanatory variables, data sources, and explanation

Explanatory variable	Data source	Brief explanation
Consumer price index (CPI)[a]	US (Bureau of Labor Statistics, 2013)	CPI measures the price level of a representative basket of goods and services purchased by urban consumers. It is one of the widely used measures of inflation
Employment level in construction (ELC)[a]	US (Bureau of Labor Statistics, 2013)	ELC is the number of employees (in thousands) on payrolls in construction
Money supply (MS)[a]	Board of Governors of the Federal Reserve Systems	MS represents the amount of money in the nation's economy. In the USA, at least two types of money supply measures are tracked: M1 and M2. M1 comprises currency in the public banks, institutions, the US treasury, travelers' checks, demand deposits, and other checkable deposits. M2 comprises M1 and savings deposits, time deposits less than $100, and balances in retail money market mutual funds. M2, which is a broader measure, was used in this research

[a]Find raw data in Appendices G, H, and I

- Examining whether the information in one time series is helpful to improve forecasting of another one (Granger causality)
- Examining whether a time series and explanatory variables are cointegrated
- Selecting a suitable multivariate model based on a cointegration test
- Forecasting the time series using the fitted model

This chapter introduces multivariate time series models and uses them to forecast HCS as an example. The required data sets can be accessed using the R package "Construction Analytics."

4.2 Potential Explanatory Time Series

Three relevant macroeconomic factors, including consumer price index (CPI), employment level in construction (ELC), and money supply (MS), are collected as potential explanatory variables for the HCS time series. Table 4.1 summarizes the potential explanatory variables for the monthly HCS time series. HCS raw data is available in Appendix G. Figure 4.1 illustrates the potential time series from January 2003 to December 2018.

4.2 Potential Explanatory Time Series

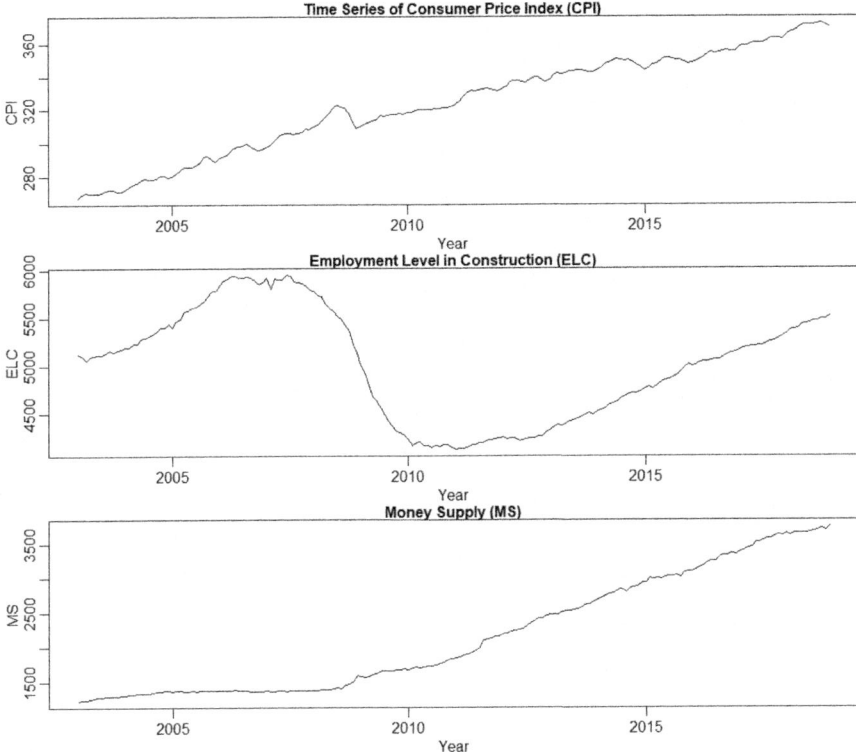

Fig. 4.1 Time series plot of CPI, ELC, and MS from January 2003 to December 2018

Table 4.2 ADF test results of CPI, ELC, and MS time series

Augmented Dickey-Fuller test
Data: cpi.ts Dickey-Fuller = −2.1412, lag order = 5, p-value = 0.5172 Alternative hypothesis: stationary
Data: elc.ts Dickey-Fuller = −2.2999, lag order = 5, p-value = 0.4509 Alternative hypothesis: stationary
Data: ms.ts Dickey-Fuller = −2.1729, lag order = 5, p-value = 0.5038 Alternative hypothesis: stationary

4.2.1 Stationarity

As the first step, a unit root test is used to identify the order of integration of the potential variables. The order of integration of the time series is defined as the number of times that a time series should be differenced to transform it into a stationary time series (Shahandashti & Ashuri, 2013). Table 4.2 shows the results of the ADF test for CPI, ELC, and MS time series. The results indicate that all the time series are nonstationary.

Table 4.3 ADF test results of differenced CPI, ELC, and MS time series

Augmented Dickey-Fuller test
Data: cpi_diff Dickey-Fuller = −6.9718, lag order = 5, p-value = 0.01 Alternative hypothesis: stationary
Data: elc_diff Dickey-Fuller = −1.9317, lag order = 5, p-value = 0.6048 Alternative hypothesis: stationary
Data: ms_diff Dickey-Fuller = −4.2585, lag order = 5, p-value = 0.01 Alternative hypothesis: stationary

Table 4.3 presents the results of the ADF test for differenced CPI, ELC, and MS time series. Consumer price index and money supply become stationary at a 1% level of significance by applying the differencing operator once except for employment level. Therefore, based on the definition, HCS, consumer price index, and money supply are integrated of order 1.

4.2.2 Granger Causality

A Granger causality test is used to test if the first differenced time series of consumer price index and money supply Granger cause the first differenced time series of HCS. The Granger causality test, as a statistical hypothesis test, is used to test if the time series of a variable is beneficial for forecasting the time series of another variable (Granger, 1969). The null hypothesis of this test is that the information in the past p-values of x is not helpful in forecasting y, where p is denoted as the lag length in the Granger causality test. Rejection of the Granger causality test's null hypothesis means that there is enough evidence to conclude that the past p-values of x can help forecast the values of y. It should be noted that the Granger causality test is sensitive to the selected lag length (p). Therefore, inspecting whether the potential explanatory variables Granger cause a time series in different lag lengths is recommended. A variable is used as an explanatory time series if the null hypothesis of the Granger causality test is rejected in at least one of the stated lag lengths. Also, a variable is a consistent explanatory variable for a time series if the null hypothesis is rejected at all stated lag lengths. R Code 4.1 tests whether ELC and MS time series Granger cause HCS time series for lag lengths 3, 6, 9, and 12. Table 4.4 summarizes the results of Granger causality for the potential explanatory variables of HCS in different lag lengths.

4.2 Potential Explanatory Time Series

R Code 4.1
Granger causality test between HCS time series and explanatory time series

```
library(zoo)
library(lmtest)
hcs_train.ts <- ts(hcs, start = c(2003,1), end = c(2017,12), frequency=12)
cpi_train.ts <- ts(cpi, start = c(2003,1), end = c(2017,12), frequency=12)
ms_train.ts <- ts(ms, start = c(2003,1), end = c(2017,12), frequency=12)
hcs_train_diff <- diff(hcs_train.ts)
cpi_train_diff <- diff(cpi_train.ts)
ms_train_diff <- diff(ms_train.ts)
hcs_train_diff <- diff(hcs_train)
grangertest(cpi_train_diff, hcs_train_diff, order = 3)
grangertest(cpi_train_diff, hcs_train_diff, order = 6)
grangertest(cpi_train_diff, hcs_train_diff, order = 9)
grangertest(cpi_train_diff, hcs_train_diff, order = 12)
grangertest(ms_train_diff, hcs_train_diff, order = 3)
grangertest(ms_train_diff, hcs_train_diff, order = 6)
grangertest(ms_train_diff, hcs_train_diff, order = 9)
grangertest(ms_train_diff, hcs_train_diff, order = 12)
```

Table 4.4 Results of Granger causality test between HCS and explanatory variables

Null hypothesis	F-statistics			
	Lag 3	Lag 6	Lag 9	Lag 12
ΔCPI does not Granger cause ΔHCS	3.2746[a]	1.49	1.8181[b]	0.7215
ΔMS does not Granger cause ΔHCS	1.6398	1.0396	0.8861	0.4702

Note: Delta (Δ) indicates the first difference operator
[a]Rejection of the null hypothesis at the 1% significance level
[b]Rejection of the null hypothesis at the 5% significance level

The results show that the first differenced consumer price index time series Granger causes the first differenced HCS, while the first differenced money supply does not Granger cause the first differenced HCS.

4.2.3 Cointegration Test

Cointegrated variables share the same stochastic trend, so they move together and do not drift apart in the long run (Wang, 2016). It is essential to use a cointegration test to examine whether a time series and its explanatory variables are cointegrated (Pfaff, 2008). Two variables with a specific order of integration are cointegrated if there is a linear combination of them with a lower order of integration, which means they are related in the long run. Engle-Granger test, Phillips-Ouliaris test, and Johansen test are popular cointegration tests commonly used to identify stable, long-run relationships between sets of variables. The extended version of the Johansen test (Johansen, 1988) (i.e., Johansen & Juselius, 1990) is used to examine if the HCS time series is cointegrated with the consumer price index and money supply time series as the potential explanatory variables.

Johansen cointegration test can be implemented using either trace or eigenvalue (Maddala & Kim, 1998). The null hypothesis of the Johansen test with the trace is that the number of cointegration vectors, r, is less than k versus the alternative, in which the number of cointegration vectors, r, is equal to k. The test proceeds sequentially for the number of cointegrations starting from 1, and the first non-rejection is used as an estimator for r. Johansen cointegration test with trace can be implemented using R Code 4.2. Table 4.5 shows the results of the Johansen cointegration test for the vector of HCS and CPI.

R Code 4.2
Johansen cointegration test

```
library(urca)
jotest=ca.jo(data.frame(hcs_train.ts,cpi_train.ts), type= "trace", ecdet="none", spec="longrun")
summary(jotest)
```

The trace statistics indicate that the null hypothesis of $r \leq 1$ is not rejected at the 5% level of significance. Therefore, HCS and the explanatory variable are cointegrated with only one relationship at a 5% significance level.

Table 4.5 Results of Johansen cointegration test for HCS and CPI time series

Null hypothesis	Trace statistics	5% critical value	1% critical value
$r \leq 1$	0.94	17.95	23.52
$r = 0$	31.85	31.52	37.22

Note: r represents the number of cointegrating relationships

4.3 Multivariate Forecasting Models

Vector autoregressive (VAR) and vector error correction (VEC) are the most common multivariate time series forecasting models. A VAR model is recommended for multivariate time series forecasting when the time series is not cointegrated with its explanatory time series. On the other hand, the VEC model is suggested for multivariate time series modeling when the vector of variables is cointegrated (Pfaff, 2008).

4.3.1 Vector Autoregressive (VAR)

A VAR model is the generalized version of the univariate autoregressive model to forecast a time series using a collection of variables. A VAR model is one of the most elastic and easy-to-use multivariate models for forecasting (Zivot & Wang, 2006). VAR models are beneficial for describing the behavior of economic and financial time series. The performance of VAR models can be superior in forecasting a construction time series compared to univariate time series models (Abediniangerabi et al., 2017). VAR models are flexible since they can be designed conditional on the potential future paths of selected variables of the model (Zivot & Wang, 2006).

VAR models are multi-equation systems in which all the variables are treated as endogenous (e.g., dependent). Each variable is represented by one equation. The general form of a VAR(p) process can be defined as:

$$Y_t = a + A_1 Y_{t-1} + A_2 Y_{t-2} + \cdots + A_p Y_{t-p} + \varepsilon_t \quad (4.1)$$

where Y_t is a vector ($n \times 1$) of time series variables, a is a vector of intercepts ($n \times 1$), A_i is a matrix ($n \times n$) of coefficient, and ε_t is a vector ($n \times 1$) of the unobservable error term (white noise).

4.3.2 Vector Error Correction (VEC)

VEC models are the restricted version of the VAR models designed for the cointegrated nonstationary variables. A VEC model limits the long-run behavior of the endogenous variables to assure their convergence to their cointegrating relationships. In the meantime, they allow for short-run adjustments. If two time series are cointegrated, there will be an error representative, which gives the inference that any changes in the dependent variable are a function of the imbalance in a cointegration relationship (represented by the error correction term) with other explanatory variables (Emmy et al., 2009).

VEC models are widely used for forecasting economic and construction time series that are nonstationary with long-run relationships by the other time series (Kim et al., 2021; Obayelu & Salau, 2010). A VEC model can help to forecast a time series, which are first differenced stationary and cointegrated (Hendry & Juselius, 2000).

The long-run form of a VEC model can be expressed by:

$$\Delta y_t = \sum_{p-1}^{i=1} A_i \Delta y_{t-i} + B y_{t-p} + C + \varepsilon t \tag{4.2}$$

where y_t is a vector ($n \times 1$) of time series at period t, n is the number of time series, A_i is a coefficient matrix ($n \times n$) for endogenous time series indicating short-term adjustments among variables, B is a cointegrating vector ($n \times n$), C is a vector ($n \times 1$) of constants, and εt is a vector ($n \times 1$) of error terms.

Since the HCS and its explanatory variable, CPI, are cointegrated, a VEC model is trained to forecast HCS for 2018. As Granger causality and cointegration tests suggest, a VEC model is developed based on HCS and CPI time series. Table 4.6 summarizes the VEC model's variables. It also includes the number of cointegrating relationships.

The "VARselect" function can select the appropriate lag lengths for developing VEC models using the explanatory time series (i.e., CPI and MS). This function provides information criteria, such as Akaike information criterion (AIC), Schwarz-Bayes (SC or BIC), Hannan-Quinn (HQ), and final prediction error (FPE), for sequentially increasing the lag order up to a VAR(p)-process (Pfaff, 2008). For a VEC model, one lag less than the lag length provided by the VARselect function can be used. R Code 4.3 shows how to use VARselect to choose a suitable lag length for the corresponding VEC models. Table 4.7 provides the results of VARselect for the proposed models.

Table 4.6 VEC models, variables, and numbers of cointegrating relationships

VEC model	Variables	r
hcs_vec	HCS, CPI	1

Note: r represents the number of cointegrating relationships

R Code 4.3
Lag selection for the proposed VEC models

```
library(vars)
VARselect(data.frame(hcs_train_diff,cpi_train_diff))
```

4.3 Multivariate Forecasting Models

Table 4.7 Recommended lag length for the proposed models based on AIC

Item	Variables	Recommended lag length	AIC
hcs_vec	HCS & CPI	9	6.120466

R Code 4.4 is used to develop the VEC model for the cointegrated vectors, including HCS and CPI, based on the recommended lag lengths.

R Code 4.4
VEC models

```
library(tsDyn)
hcs_vec <- VECM(data.frame(hcs_train.ts, cpi_train.ts), lag=9)
summary(hcs_vec)
```

R Code 4.5 forecasts HCS for 12 months of 2018 using the trained VEC model. Table 4.8 shows the forecasts for HCS for 2018 using the trained VEC and SARIMA models developed in Chap. 2. Figure 4.2 illustrates the HCS time series from 2003 to 2018 and its forecasts for 2018.

R Code 4.5
HCS forecasts using VEC models

```
hcs_vec_prediction = predict(hcs_vec, n.ahead = 12)
```

4.3.3 Forecasting Errors of VEC Models

The performance of the trained VEC model is compared with that of the univariate SARIMA model (see Chap. 2) in forecasting HCS time series. The comparison is based on four error measures: mean absolute prediction error (MAPE), mean square error (MSE), root-mean-square error (RMSE) and mean absolute error (MAE). These error measures are expressed as follows:

$$\text{MAPE} = \frac{1}{n}\sum_{n}^{i=1} \frac{\left|\hat{Y} - Y_t\right|}{Y_t} \tag{4.3}$$

$$\text{MSE} = \frac{1}{n}\sum_{n}^{i=1} \left(\hat{Y}_t - Y_t\right)^2 \tag{4.4}$$

Table 4.8 HCS forecasts for 12 months of 2018 using VEC models

Point	Actual HCS	hcs_vec	SARIMA(0,1,2)(0,1,3)$_{12}$
Jan 2018	63	69.03825	64.64307
Feb 2018	60	67.52146	57.89494
Mar 2018	67	66.7669	68.96394
Apr 2018	68	68.32807	77.29724
May 2018	81	80.11315	86.63034
Jun 2018	82	83.00978	99.85132
Jul 2018	70	103.90838	117.51224
Aug 2018	102	102.71754	112.46465
Sep 2018	152	101.64608	135.29342
Oct 2018	101	90.14993	98.00904
Nov 2018	94	83.0063	81.95908
Dec 2018	75	71.95671	63.9463

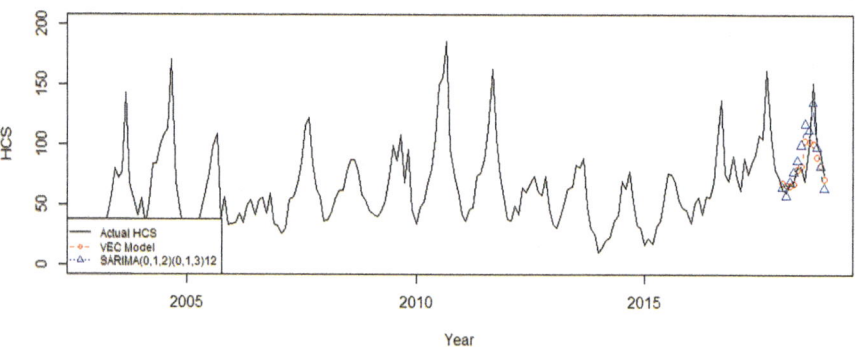

Fig. 4.2 HCS time series forecasts for the year 2018 using vector error correction and SARIMA models

$$\text{RMSE} = \sqrt{\frac{1}{n}\sum_{n}^{i=1}\left(\hat{Y}_t - Y_t\right)^2} \quad (4.5)$$

$$\text{MAE} = \frac{1}{n}\sum_{n}^{i=1}\left|\hat{Y} - Y_t\right| \quad (4.6)$$

where $\hat{Y}_t$ is the forecasted HCS, Y_t is the actual HCS, and n is denoted as the total number of forecasted data points. Table 4.9 presents the MAPE, MSE, RMSE, and MAE metrics measured using forecasted data points from the models from January 2018 to December 2018. The error metrics show that the bivariate VEC model provides lower MAPE and MAE compared to the univariate SARIMA model.

4.5 Exercise Problems

Table 4.9 Forecasting errors of time series models

Point	hcs_vec	SARIMA
MAPE (%)	11.17	14.25
MSE	167.8608	140.4143
RMSE	12.95611	11.84965
MAE	10.49037	11.605

4.4 Summary

This chapter presents multivariate time series forecasting models to predict construction time series. It explains how to identify a leading indicator of the time series model and use it to improve the accuracy of construction forecasting. Several practical examples and R codes are provided to create and diagnose two multivariate time series models, such as vector error correction and vector autoregressive models for forecasting construction variables. The comparison results indicate that the multivariate time series models can outperform the univariate time series models for predicting construction variables.

4.5 Exercise Problems

Use the time series provided in the appendices (i.e., J, K, L, and M) to answer the problems.

1. Apply ADF test on Sim1 and Sim2 time series and determine whether they are stationary time series. If not, use the diff operator to make them stationary.
2. Apply Granger causality to test to determine whether Sim2 Granger causes Sim1 time series for lag lengths 3, 6, 9, and 12.
3. Use the Johansen cointegration test to determine the number of cointegration relationships between the Sim1 and Sim2 time series.
4. Based on the results of previous questions, which multivariate forecasting model is appropriate for forecasting Sim1 time series?
5. Test whether the US crude oil price (COP) time series helps predict the California Construction Cost Index (CCCI). Are the two time series cointegrated in the long run?
6. Develop a bivariate VEC model for the California Construction Cost Index (CCCI) using the crude oil price (COP) time series.

References

Abediniangerabi, B., Shahandashti, S. M., Ahmadi, N., & Ashuri, B. (2017). Empirical investigation of temporal association between architecture billings index and construction spending using time-series methods. *Journal of Construction Engineering and Management, 143*(10), 04017080.

Bureau of Labor Statistics (BLS). (2013). http://www.bls.gov/ppi/. June 7, 2013.

Emmy, F. A., Baharom, A. H., Radam, A., & Illisriyani, I. (2009). *Export and import cointegration in forestry domain: The case of Malaysia* (MPRA Paper No. 16673). MPRA.

Granger, C. W. (1969). Investigating causal relations by econometric models and cross-spectral methods. *Econometrica: Journal of the Econometric Society, 37*, 424–438.

Hendry, D. F., & Juselius, K. (2000). Explaining cointegration analysis: Part 1. *The Energy Journal, 21*, 1–42.

Johansen, S. (1988). Statistical analysis of cointegration vectors. *Journal of Economic Dynamics and Control, 12*(2–3), 231–254.

Johansen, S., & Juselius, K. (1990). Maximum likelihood estimation and inference on cointegration—With applications to the demand for money. *Oxford Bulletin of Economics and Statistics, 52*(2), 169–210.

Kim, S., Abediniangerabi, B., & Shahandashti, M. (2021). Pipeline construction cost forecasting using multivariate time series methods. *Journal of Pipeline Systems Engineering and Practice, 12*(3), 04021026.

Maddala, G. S., & Kim, I. M. (1998). *Unit roots, cointegration, and structural change*. Cambridge University Press.

Obayelu, A. E., & Salau, A. S. (2010). Agricultural response to prices and exchange rate in Nigeria: Application of co-integration and vector error correction model (VECM). *Journal of Agricultural Sciences, 1*(2), 73–81.

Pfaff, B. (2008). *Analysis of integrated and cointegrated time series with R*. Springer Science & Business Media.

Shahandashti, S. M., & Ashuri, B. (2013). Forecasting engineering news-record construction cost index using multivariate time series models. *Journal of Construction Engineering and Management, 139*(9), 1237–1243.

Wang, W. (2016). *Achieving inclusive growth in China through vertical specialization*. Chandos Publishing.

Zivot, E., & Wang, J. (2006). Vector autoregressive models for multivariate time series. In *Modeling financial time series with S-Plus®* (pp. 385–429). Springer.

Chapter 5
Construction Forecasting Using Recurrent Neural Networks

Abstract Despite all their advantages, univariate and multivariate time series models are linear statistical methods subject to significant limitations for characterizing nonlinear relationships. Machine learning models, such as neural networks, have established themselves as a serious alternative to classical statistical models for exploring nonlinear relationships. This chapter introduces recurrent neural networks (e.g., long short-term memory and gated recurrent units) for nonlinear time series forecasting and explains the life cycle of construction time series forecasting using such networks. These networks are designed and trained to forecast Highway Construction Spending (HCS) time series. Also, their forecasting performances are investigated and compared with those of seasonal ARIMA and VEC models. The comparison results show that recurrent neural networks (i.e., long short-term memory and gated recurrent unit networks) can provide higher accuracies in forecasting the long-term variations of HCS than statistical linear time series models based on typical error measures.

Keywords Construction cost forecasting · Machine learning methods · Recurrent neural networks · Long short-term memory · Gated recurrent unit

5.1 Introduction

Linear statistical methods (e.g., autoregressive moving average (ARMA) model) have been successfully used in many construction applications (Abediniangerabi et al., 2017). However, their linear assumption limits their application for some construction applications that require exploring nonlinear relationships (Bontempi et al., 2012). Recent advancements in big data acquisition and management in architecture, engineering, and construction (AEC) industry have resulted in time series that require exploring nonlinear relationships to investigate chronic construction problems. Machine learning models have established themselves as a serious alternative to classical statistical models for exploring nonlinear and nonparametric relationships. Machine learning models are often black box or data driven (Mitchell, 1997). Among various machine learning algorithms, artificial neural networks

(ANNs) have received substantial consideration from the scientific community in forecasting time series (Tealab, 2018). ANN structures are similar to biological brain cells, which process information for humans. A neuron is a nonlinear function mapping inputs to outputs (Haykin, 2010):

$$o = f\left(x_1, x_2, \ldots, x_n; w_1, \ldots, w_2, \ldots, w_p\right) = f(x, w)$$

where x is the input vector of variables into a neuron, w is the weights (parameters) associated with the inputs of the neuron, and $f(.)$ is a nonlinear activation function. Generally speaking, an artificial neural network (ANN) is a composition of nonlinear functions. ANNs comprise of basic structure in which small processing units are connected by weighted connections (Graves, 2012). Figure 5.1 illustrates a simple ANN. In this ANN, nodes and connections represent the neurons and the synapses between the neurons, respectively. A neural network is trained based on the input data where the weight of each node in the network is determined and then optimized to model the mapping in the sets of input/output pairs. A network can forecast the outputs corresponding to responses that it never "saw" once it is trained.

Generally, there are two types of artificial neural networks (i.e., feed-forward networks and feedback (recurrent) networks). The feed-forward networks (FNNs), also called static networks, pass the data forward from input to output, whereas recurrent neural networks (RNNs) feed the data back into the input layer where it is fed forward again to be further processed and generate the final output using a feedback loop in its structure (Dematos et al., 1996). Figure 5.1 clearly shows how the data enter from the input layer into the network and are passed to the output layer within an FNN with no cycle. On the other hand, a feedback network, known as the recurrent network, is characterized by cycles in a way that the outputs of the neurons in a layer can be looped back to the inputs to the same neuron or inputs to the previous layers' neurons. Figure 5.2 illustrates the structure of a feedback neural network.

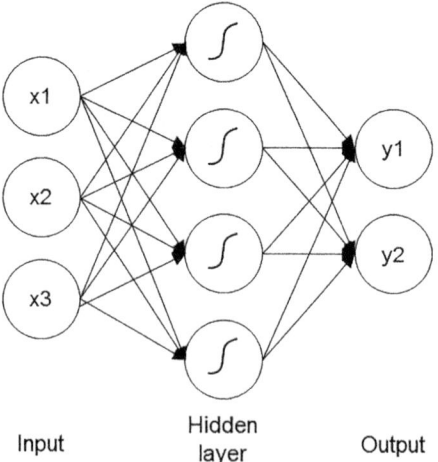

Fig. 5.1 Simple multilayer perceptron with three inputs

5.2 Recurrent Neural Networks

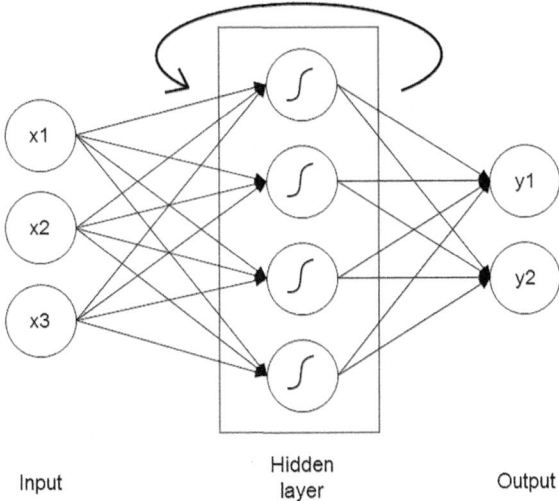

Fig. 5.2 A feedback neural network with one hidden layer

5.2 Recurrent Neural Networks

A recurrent neural network (RNN) is an extended version of conventional feed-forward neural networks with cycles to handle a variable-length sequence of inputs (Chung et al., 2014). Figure 5.3 illustrates a simple recurrent neural unit (folded and unfolded) with three inputs. A feed-forward neural network can only map the data in one direction from input vector to output vector; in contrast, an RNN can map the entire history of inputs from previous lags to each output by introducing loops in the network (Graves, 2012). In an RNN, the recurrent connections allow a "memory" of inputs from previous lags to endure in the network's internal state, which can be used to impact the network outputs (Goodfellow et al., 2016). The memory helps the network to learn long-term dependencies in a sequence. For instance, the input X at time $t-1$ can be considered in the prediction of output value at times $t-1$ as well as values at t, $t+1$, and so on. This capability of RNNs makes them powerful sequence learners well suited for time series analysis (Graves, 2012).

In this chapter, three recurrent neural networks, including simple recurrent neural networks (RNNs), long short-term memory (LSTM) networks, and gated recurrent unit (GRU) networks, are briefly explained and used for forecasting Highway Construction Spending (HCS) time series (Appendix F). TensorFlow and Keras packages are required for implementing RNNs. R Code 5.1 can be used to install these packages on RStudio.

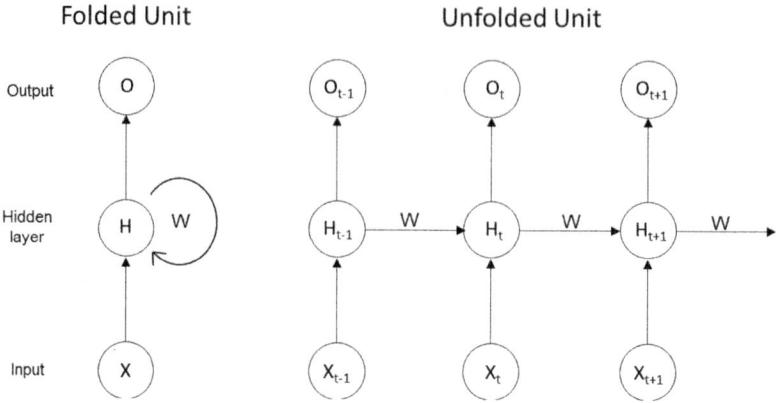

Fig. 5.3 A simple recurrent neural unit (folded and unfolded)

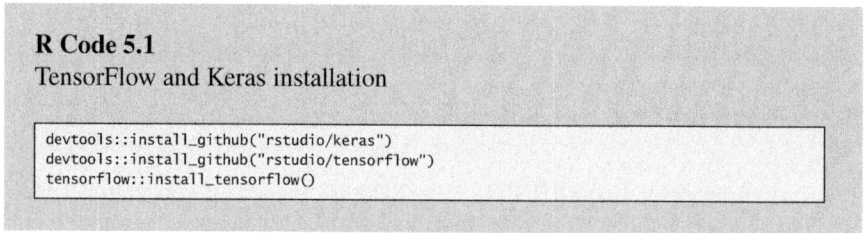

R Code 5.1
TensorFlow and Keras installation

```
devtools::install_github("rstudio/keras")
devtools::install_github("rstudio/tensorflow")
tensorflow::install_tensorflow()
```

5.3 Model Development Process

The following subsections provide a walk-through of implementing recurrent neural networks for forecasting construction time series.

5.3.1 Data Preparation

RNNs are supervised machine learning models. Both target and explanatory variables are required to be specified. A k-step lagged dataset can be generated by transforming the construction time series by lagging the time series and having the input value at the time $(t - k)$ and a target value at the time t. For example, R Code 5.2 generates a k-step lagged dataset from the HCS time series.

5.3 Model Development Process

R Code 5.2
Lag transformation

```r
#Creating a transformation function
lag_transform <- function(x, k = 1){
  lagged = c(rep(NA, k), x[1:(length(x)-k)])
  DF = as.data.frame(cbind(lagged, x))
  colnames(DF) <- c( paste0('x-', k), 'x')
  DF[is.na(DF)] <- 0
  return(DF)
}
Supervised_hcs = lag_transform(hcs.ts, 1) #Transformation HCS to 1-step lagged time series
head(Supervised_hcs)
```

5.3.2 Train and Test Datasets

The first lagged dataset is split into training and testing datasets. The training dataset is used for training the network, and the testing dataset is preserved to evaluate the testing performance of the trained network on an unseen dataset. It seems that there is no clear guidance on the train/test ratio for a given dataset (Joseph, 2022). Based on the Pareto principle, the typical train/test split ratio is 80/20 (Hemdan et al., 2020). However, other ratios, such as 90:10, 70:30, and even 50:50 are also used in practice (Joseph, 2022). HCS time series includes monthly data points for 16 years from 2003 to 2018. One year of monthly data (i.e., 12 data points) is used for testing a recurrent neural network for short-term forecasting. R Code 5.3 can be used to split the HCS time series into training and testing datasets while conserving the order of observations. There are 180 observations (93.75% of the observations) used for training the network, and 12 observations (6.25% of the observations) are kept as the out-of-sample testing dataset.

R Code 5.3
Splitting HCS time series

```r
N = nrow(Supervised_hcs) # Finding the length of the time series
n = round(N *0.9375, digits = 0)   # 12 months
train = Supervised_hcs[1:n, ] # Slicing the train dataset
test = Supervised_hcs[(n+1):N, ] # Slicing the test dataset
```

5.3.3 Data Rescaling

ANNs are sensitive to the scale of the data. Moreover, rescaling the data can speed up the model training process. Therefore, it is recommended to scale the data to the range of 0 to 1 before training RNNs (Goodfellow et al., 2016). R Code 5.4

implements the min-max scaling function. The "*scale_data*" function is used to simultaneously scale the training and testing sets of a time series. This function finds the minimum and maximum values in the time series and transforms them into 0 and 1, and other values in the time series are transformed into a decimal between 0 and 1 (Eq. 5.1).

$$x' = \frac{x - \min(x)}{\max(x) - \min(x)} \tag{5.1}$$

R Code 5.4
Min-max scaling function

```
scale_data = function(train, test, feature_range = c(0, 1)) {
  x = train
  fr_min = feature_range[1]
  fr_max = feature_range[2]
  std_train = ((x - min(x) ) / (max(x) - min(x) ))
  std_test = ((test - min(x) ) / (max(x) - min(x) ))
  scaled_train = std_train *(fr_max -fr_min) + fr_min
  scaled_test = std_test *(fr_max -fr_min) + fr_min
  return( list(scaled_train = as.vector(scaled_train), scaled_test = as.vector(scaled_test)
  ,scaler= c(min =min(x), max = max(x))) )
}
```

R Code 5.5 implements the inverse rescaling function. The "*invert_scaling*" can be used after predicting to rescale back the predicted time series into the original scale. R Code 5.6. uses the rescale function to rescale HCS time series training and testing datasets within the range of 0 and 1.

5.3.4 RNN Parameters

Similar to any machine learning algorithm, RNNs have several hyperparameters that need to be specified for training to achieve higher prediction performance. A model parameter should be estimated from data. For example, the weight coefficients of a linear regression model are model parameters. On the other hand, a hyperparameter is a parameter that controls the learning process within a model. For instance, the number of hidden layers in an RNN model is a hyperparameter.

5.3 Model Development Process

R Code 5.5
Min-max inverse rescaling function

```
invert_scaling = function(scaled, scaler, feature_range = c(0, 1)){
    min = scaler[1]
    max = scaler[2]
    t = length(scaled)
    mins = feature_range[1]
    maxs = feature_range[2]
    inverted_dfs = numeric(t)
    for( i in 1:t){
        X = (scaled[i]- mins)/(maxs - mins)
        rawValues = X *(max - min) + min
        inverted_dfs[i] <- rawValues
    }
    return(inverted_dfs)
}
```

R Code 5.6
Rescaling HCS time series train and test sets

```
Scaled = scale_data(train, test, c(0, 1))
y_train = Scaled$scaled_train[, 2]
x_train = Scaled$scaled_train[, 1]
y_test = Scaled$scaled_test[, 2]
x_test = Scaled$scaled_test[, 1]
```

The main hyperparameter of RNNs is the structure of the network. The number of hidden layers, types of hidden layers, and the number of units in each layer shape the structure of a recurrent neural network. Dropout layers, weight initialization, activation function, learning rate, momentum, number of epochs, and batch size are other hyperparameters that require to be tuned to improve the prediction performance of RNNs. There is often no rule for selecting these hyperparameters, and most of the time, it is impossible to immediately define the best hyperparameters for a given model. Hyperparameter tuning helps to find the optimum values for these hyperparameters for each problem by exploring a range of possibilities. Hyperparameter tuning is defined as the task of choosing a set of optimal hyperparameters for a learning algorithm.

Neural networks are trained through several epochs (i.e., iterations). Epoch number (i.e., the number of iterations) shows how many times the entire dataset is passed

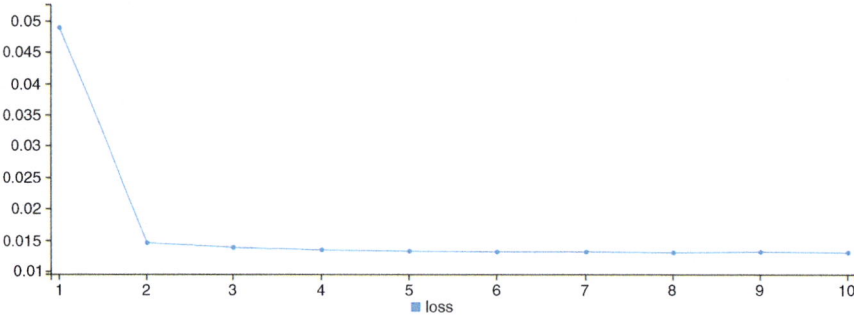

Fig. 5.4 An example of loss function saturation through several epochs

forward and backward through the network to update the weights of the network. Usually, a network cannot be trained well with only 1 epoch. A larger epoch number can assure preventing underfitting. On the other hand, more epochs for training require more computation power. It may also lead to overfitting. It is recommended to set the initial epoch number high enough to saturate the loss function during the training process. Figure 5.4. illustrates an example of the loss function for a network through 10 epochs. This figure clearly shows that the loss function starts to saturate after 4 epochs; therefore, an epoch number of 5 is adequate to train the network.

5.4 Simple RNNs

The "Keras" library is used to construct a sequential network. Keras is a powerful and open-source library for developing and evaluating deep learning models. In simple RNNs, dense layers are used as hidden layers. A dense layer is defined as a layer of neurons in a neural network that receives the input data from all the neurons of the previous layer, so it is densely connected.

A simple recurrent neural network was developed using a sequential model and trained on the HCS training dataset (e.g., monthly HCS time series from 2003 to 2017) to forecast the HCS time series for 2018 (Fig. 5.5). The network has three dense layers. The first and second layers have 12 and 6 units, respectively. The last dense layer is a fully connected layer to generate the target values. The rectified linear unit (ReLU) was used as the activation function in the first and second dense layers (Eq. 5.2).

$$f(x) = (0,x) \qquad (5.2)$$

For any negative input, the ReLU activation functions will return 0; on the other hand, for any positive value (e.g., x), the function will return that value. Therefore, the output has a range of 0 to infinite. RMSprop optimizer was used to minimize the loss function (i.e., mean squared error (MSE)) of the network. R Code 5.7 was developed to specify the model hyperparameters, fit the model, and forecast the HCS time series for 12 months.

5.4 Simple RNNs

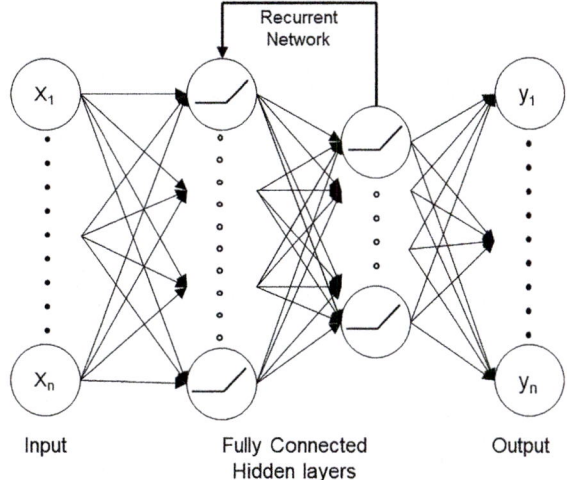

Fig. 5.5 Structure of a simple RNN network developed to forecast the HCS time series

R Code 5.7
Simple RNN

```r
#Call the library
library(keras)
#Set seed to be able to get reproducible results
set_random_seed(1)
#Reshape the data
dim(x_train) <- c(length(x_train), 1, 1)
dim(y_train) <- c(length(y_train), 1, 1)
X_shape2 = dim(x_train)[2]
X_shape3 = dim(x_train)[3]
#Define the batch size
batch_size = 1
#Define Model Structure
model_rnn <- keras_model_sequential() %>%
  layer_dense(units = 12, batch_input_shape = c(batch_size, X_shape2, X_shape3), activation = "relu") %>%
  layer_dense(units = 6, activation = "relu") %>%
  layer_dense(units = 1)
#Compile the model
model_rnn %>% compile(optimizer = optimizer_rmsprop(),loss = "mean_squared_error")
#Define the epoch number
epochs = 10
#Train the model
for(i in 1:epochs){
  model_rnn %>% fit(x_train, y_train, batch_size = batch_size, verbose = 1,shuffle = FALSE)
  model_rnn %>% reset_states()
}
L = length(x_test)
scaler = Scaled$scaler
predictions_rnn = numeric(L)
#Test the model
for(i in 1:L){
  X = x_test[i]
  dim(X) = c(1,1,1)
  yhat = model_rnn %>% predict(X, batch_size=batch_size)
  #Inverst the predictions
  yhat = invert_scaling(yhat, scaler, c(0, 1))
  predictions_rnn[i] <- yhat}
#Print the predictions
print(predictions_rnn)
predictions_rnn_ts <- ts(predictions_rnn, start = c(2018,1),frequency = 12)
#Calcutlate the prediction performance metrics
accuracy(predictions_rnn_ts, test$x)
```

Results

```
[1]  75.39227  66.72186  64.35721  69.87474  70.66296  80.90980  81.69802  72.23940  96.13
637 128.77708  95.48356  90.71832

                  ME       RMSE       MAE       MPE      MAPE
Test set  1.835702  21.37251  14.78001  -1.367888  15.36209
```

Table 5.1 HCS's actual values and forecasts for the 12 months of 2018 using a three-layer RNN

Point	Actual values	RNN forecasts
Jan 2018	63	75.39227
Feb 2018	60	66.72186
Mar 2018	67	64.35721
Apr 2018	68	69.87474
May 2018	81	70.66296
Jun 2018	82	80.9098
Jul 2018	70	81.69802
Aug 2018	102	72.2394
Sep 2018	152	96.13637
Oct 2018	101	128.77708
Nov 2018	94	95.48356
Dec 2018	75	90.71832

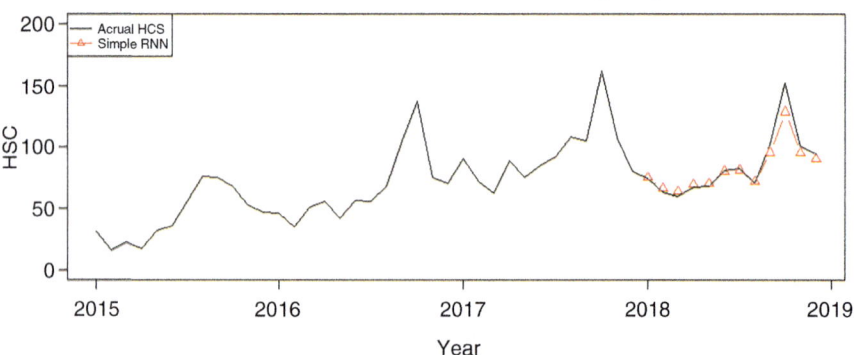

Fig. 5.6 HCS time series forecasts for 2018 using a simple RNN with three hidden layers

Table 5.1 presents the forecasts for the HCS time series for the 12 months of 2018. Figure 5.6 illustrates the HCS time series forecasts for this period. The figure shows that the network can effectively model the trend and seasonality in the data.

5.5 Long Short-Term Memory (LSTM)

Recurrent neural networks use past information during the training to map the input sequences to the output sequences. However, access to past information is limited in a simple RNN due to the common problem of artificial neural networks known as

5.5 Long Short-Term Memory (LSTM)

vanishing or exploding gradients. In each iteration in the training phase, the neural network's weights are updated proportionally to the partial derivative of the error function concerning the current weights in each training iteration. Sometimes, these partial derivatives are too small (e.g., numbers between -1 and 1), hindering the weight from changing its value. Since these values are small, multiplying these numbers will result in decreases in a gradient, which is called the vanishing gradient problem. Different activation functions that their derivatives can take larger values may be used. However, these activation functions can also result in exponential increases in the gradient, called exploding gradient descent (Graves, 2012).

One difference between ANNs and RNNs is that, in RNNs, the recurring weights should be propagated all the way back through time to the neurons in the hidden layers during each update. The recurring weights are used to connect the hidden layers to themselves in the unrolled temporal loop. During the training phase of an RNN, the cost function compares the outputs of the model with the desired outputs (e.g., ground truth). Then, the cost function is calculated and propagated back through the network to update the weights in the RNN. Every single neuron that participated in the calculation of output should update its weights to minimize the error. But these weights are too small and close to zero. Therefore, the gradient values are shrinking exponentially with small values in the matrix. Multiple matrix multiplications with these small values eventually vanish the magnitudes of values after a few time steps. As a result, gradient contributions of "faraway" steps now become zero, and the state at those steps does not contribute to what is being learned. The model ends up not learning long-range dependencies. Figure 5.7 illustrates an example of a vanishing gradient problem in a simple RNN through time. In this example, the figure shows that the contribution of each input in forecasting the output is fading over time. For instance, the figure shows that the input data at time 1 does not contribute to the calculation of the output at time 6 due to the vanishing problem in RNNs, while the input values at times 2, 3, 4, and 5 contribute to the output at time 6.

Several solutions have been recommended to resolve the vanishing or exploding gradients in RNNs. One solution is using long short-term memory (LSTM) units in the RNNs. The LSTM layer is a specific type of hidden layer that solves the vanishing gradient problem by introducing memory blocks. Figure 5.8 shows a memory block in an LSTM layer. A memory block (or unit) is composed of a cell. This cell comprises

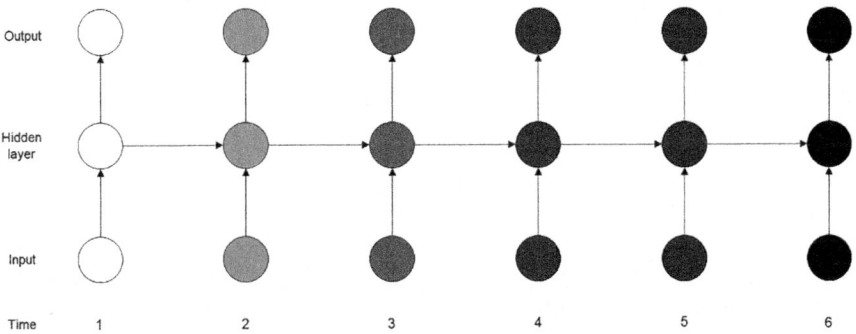

Fig. 5.7 Vanishing gradient problem in recurrent neural networks

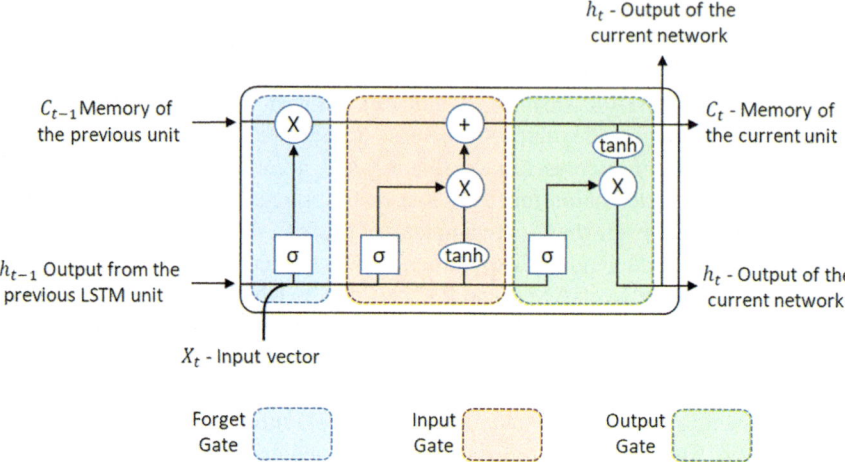

Fig. 5.8 Structure of a memory cell within an LSTM neural network

an input gate, an output gate, and a forget gate. The cell remembers values over random time intervals; the three gates control the information flow into and also out of the cell.

A cell takes three values, including X_t, C_{t-1}, and h_{t-1}, as inputs. X_t is the input of the current time step. C_{t-1} is the memory of the previous unit, and h_{t-1} is the output from the previous LSTM unit. The cell also generates two outputs: h_t and C_t, which are the output of the current network and the memory of the current unit, respectively. A single unit takes into consideration the current input, previous output, and previous memory, generating a new output. It also alters its memory with the given structure. Gates are used to optionally let information flow through a cell. They are composed of an activation function (i.e., tanh) and an element-wise operation. Each gate has a specific task within a cell. Forget gate determines what information can be detached from the cell. An input gate determines what new information can be saved in the cell. Finally, an output gate decides what information can be transferred to the next cell as an input.

Memory cells in LSTM layers enable RNNs to access, store, and alter information over long periods to prevent vanishing or exploding gradient problems. Figure 5.9 illustrates the effect of using memory cells in an LSTM layer within a neural network on preserving gradient information. As shown in the figure, cells are sensitive to input units. These memory cells can remember previous inputs (e.g., time 1) if the forget gate is open, and the input gate is closed. For example, for the given output at time 6, the input value at time 1 is used in estimating the output value at time 6. The figure also shows that the input value at time 1 is used to calculate the output values at times 3, 4, and 6, but not at times 2 and 5.

An RNN with an LSTM layer was developed and trained using monthly HCS time series from 2003 to 2017 (13 years) to forecast the HCS time series for 12 months. R Code 5.8 was used to design the LSTM model structure, fit the model, and forecast the HCS time series. The network had one LSTM layer (6 units) and a dense layer for the output. The epoch number was set to 10 to fit the model.

5.5 Long Short-Term Memory (LSTM)

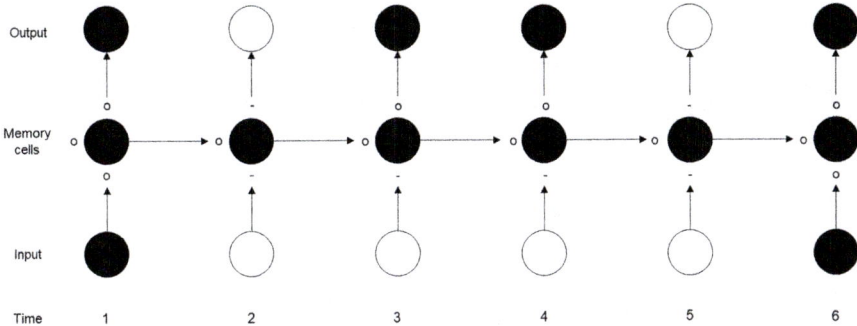

Fig. 5.9 LSTM mechanism to solve vanishing or exploding gradient problems in a recurrent neural network using memory cells

R Code 5.8
LSTM RNN specification

```r
#Call the library
library(keras)
#Set seed to be able to get reproducible results
set_random_seed(2)
#Reshape the data
dim(x_train) <- c(length(x_train), 1, 1)
X_shape2 = dim(x_train)[2]
X_shape3 = dim(x_train)[3]
#Define the batch size
batch_size = 1
#Define Model Structure
model_lstm <- keras_model_sequential()
model_lstm%>%
   layer_lstm(units=6, batch_input_shape = c(batch_size, X_shape2, X_shape3), stateful=TRUE)%>%
   layer_dense(units = 1)
#Compile the model
model_lstm %>% compile(loss = 'mean_squared_error', optimizer = optimizer_adam( lr= 0.02, decay = 1e-4 ), metrics = c('accuracy'))
#Define the epoch number
epochs = 10
#Train the model
for(i in 1:Epochs ){
   model_lstm %>% fit(x_train, y_train, epochs=2, batch_size=batch_size, verbose=1, shuffle=TRUE)
   model_lstm %>% reset_states()
}
L = length(x_test)
scaler = Scaled$scaler
predictions_lstm = numeric(L)
#Test the model
for(i in 1:L){
  X = x_test[i]
  dim(X) = c(1,1,1)
  yhat = model_lstm %>% predict(X, batch_size=batch_size)
    #Inverst the predictions
  yhat = invert_scaling(yhat, scaler, c(0, 1))
  predictions_lstm[i] <- yhat}
#Print the predictions
print(predictions_lstm)
predictions_lstm_ts <- ts(predictions_lstm, start = c(2018,1),frequency = 12)
#Calcutlate the prediction performance metrics
accuracy(predictions_lstm_ts, test$x)
```

Results

```
[1]   72.02677  61.34595  62.10513  68.33535  68.94496  79.46456  79.77966  71.40145  93.65
134  111.05433  94.97430  92.51853
              ME       RMSE       MAE       MPE      MAPE
Test set 4.949806  20.62469  13.12229  2.356794  13.25427
```

Table 5.2 HCS's actual values and forecasts for the 12 months of 2018

Point	Actual values	LSTM forecasts
Jan 2018	63	72.02677
Feb 2018	60	61.34595
Mar 2018	67	62.10513
Apr 2018	68	68.33535
May 2018	81	68.94496
Jun 2018	82	79.46456
Jul 2018	70	79.77966
Aug 2018	102	71.40145
Sep 2018	152	93.65134
Oct 2018	101	111.05433
Nov 2018	94	94.9743
Dec 2018	75	92.51853

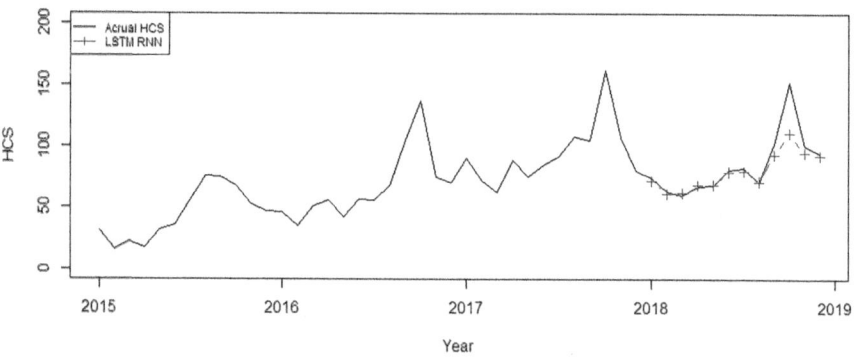

Fig. 5.10 HCS time series forecasts for 2018 using LSTM RNN

Table 5.2 presents the forecast of the HCS time series for 12 months of 2018 using the trained LSTM model. Figure 5.10 illustrates the HCS time series from 2003 to 2017 and its forecasts for 2018. Similar to simple RNN, LSTM also captures the trend and seasonality of the data.

5.6 Gated Recurrent Unit (GRU)

Another solution that has been proposed to solve the vanishing or exploding gradients in RNNs is using gated recurrent units (GRU). Each recurrent unit in GRU adaptively considers different time scales' dependencies using two gates: a reset gate r (for adjusting the incorporation of new input data with the previous memory) and an update gate z (for regulating the preservation of the information in the previous memory) (Zhao et al., 2017). Figure 5.11 illustrates a gated recurrent unit (GRU) in an RNN network. A GRU unit has similarities with an LSTM unit. Both networks have an extra component from t to $t + 1$ to add new information to the existing content. However, while LSTM preserves an internal memory state cell, GRU does not provide a separate memory cell. Therefore, GRU exposes the full hidden content without any control, which makes the training phase faster (Chung et al., 2014).

An RNN with a GRU layer was developed and trained using monthly HCS time series from 2003 to 2017 (15 years) to forecast the HCS time series for 2018. R Code 5.9 was used to design an RNN with a GRU layer with three neurons, fit the model, and forecast the HCS time series. The epoch number was set to 10 to fit the model.

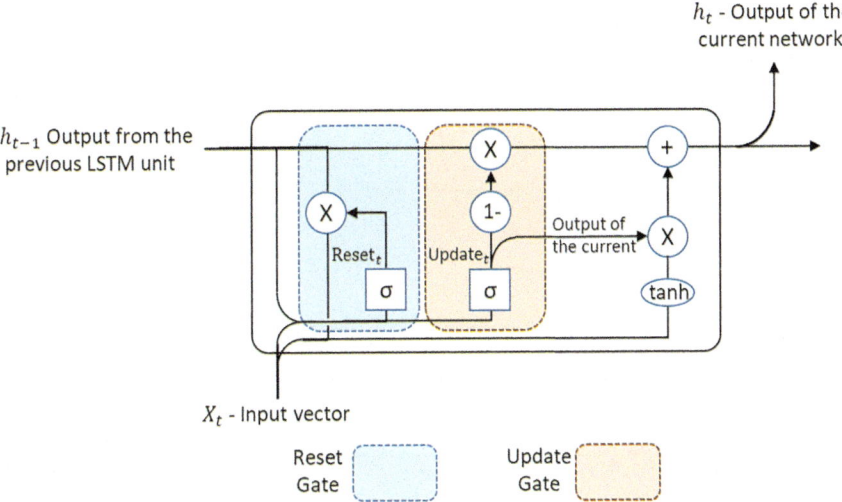

Fig. 5.11 Structure of a gated recurrent unit

R Code 5.9
GRU RNN specification

```r
#Call the library
library(keras)
#Set seed to be able to get reproducible results
set_random_sid(3)
#Reshape the data
dim(x_train) <- c(length(x_train), 1, 1)
X_shape2 = dim(x_train)[2]
X_shape3 = dim(x_train)[3]
#Define the batch size
batch_size = 10
#Define Model Structure
model_gru <- keras_model_sequential()
model_gru %>%
  layer_gru(units=3, batch_input_shape = c(batch_size, X_shape2, X_shape3)) %>%
  layer_dense(units = 1)
#Compile the model
model_gru %>% compile(optimizer = optimizer_rmsprop(),loss = "mean_squared_error")
#Define the epoch number
Epochs = 10
#Train the model
for(i in 1:Epochs ){
  model_gru %>% fit(x_train,  y_train,  epochs=1,  batch_size=batch_size, verbose=1, shuffle=FALSE)
  model_gru %>% reset_states()
}
L = length(x_test)
scaler = Scaled$scaler
predictions_gru = numeric(L)
#Test the model
for(i in 1:L){
  X = x_test[i]
  dim(X) = c(1,1,1)
  yhat = model_gru %>% predict(X, batch_size=batch_size)
  #Inverst the predictions
  yhat = invert_scaling(yhat, scaler,  c(0, 1))
  predictions_gru[i] <- yhat}
#Print the predictions
print(predictions_gru)
predictions_gru_ts <- ts(predictions_gru, start = c(2018,1),frequency = 12)
#Calcutlate the prediction performance metrics
accuracy(predictions_gru_ts, test$x)
```

Results

```
 [1]  71.22985  66.13967  64.75191  67.99080  68.45363  74.46421  74.92561  69.37924  84.095
12 106.13074  83.64001  80.44312
                  ME     RMSE      MAE      MPE    MAPE
Test set  8.613008 22.70721 13.59117 6.087903 13.19958
```

Table 5.3 presents the forecast of the HCS time series for the 12 months of 2018 using the GRU model. Figure 5.12 illustrates the HCS time series from 2003 to 2017 and its forecasts for 2018.

5.7 Forecasting Errors of Time Series Models

Table 5.3 HCS's actual values and forecasts for the 12 months of 2018

Point	Actual values	GRU forecasts
Jan 2018	63	71.22985
Feb 2018	60	66.13967
Mar 2018	67	64.75191
Apr 2018	68	67.9908
May 2018	81	68.45363
Jun 2018	82	74.46421
Jul 2018	70	74.92561
Aug 2018	102	69.37924
Sep 2018	152	84.09512
Oct 2018	101	106.13074
Nov 2018	94	83.64001
Dec 2018	75	80.44312

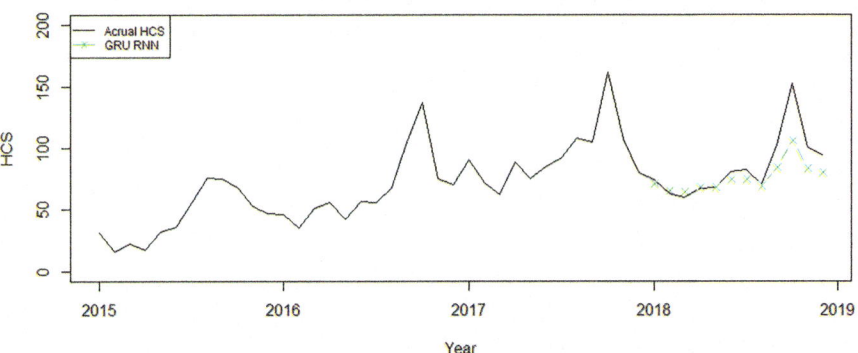

Fig. 5.12 HCS time series forecasts for 2018 using the GRU RNN

5.7 Forecasting Errors of Time Series Models

Table 5.4 and Fig. 5.13 present HCS forecasts using simple RNN, LSTM, GRU, seasonal autoregressive integrated moving average (SARIMA), and vector error correction (VEC) models altogether. The figure shows that RNNs could accurately predict the seasonality and trend of HCS through the 12 months of 2018. The SARIMA and VEC models (developed in previous chapters) could also capture the general trend and seasonality of the HCS time series but not as accurately as RNNs. The figure shows that the SARIMA and VEC models perform better in the short term (e.g., 3 months) than in the long term.

Table 5.4 HCS's actual values and forecasts for the 12 months of 2018

Point	Actual values	Simple RNN forecasts	LSTM forecasts	GRU forecasts	SARIMA(0,1,2)(0,1,3)$_{12}$ forecasts	Bivariate VEC
Jan 2018	63	75.39227	72.02677	71.22985	64.64307	69.03825
Feb 2018	60	66.72186	61.34595	66.13967	57.89494	67.52146
Mar 2018	67	64.35721	62.10513	64.75191	68.96394	66.7669
Apr 2018	68	69.87474	68.33535	67.9908	77.29724	68.32807
May 2018	81	70.66296	68.94496	68.45363	86.63034	80.11315
Jun 2018	82	80.9098	79.46456	74.46421	99.85132	83.00978
Jul 2018	70	81.69802	79.77966	74.92561	117.51224	103.9084
Aug 2018	102	72.2394	71.40145	69.37924	112.46465	102.7175
Sep 2018	152	96.13637	93.65134	84.09512	135.29342	101.6461
Oct 2018	101	128.77708	111.05433	106.13074	98.00904	90.14993
Nov 2018	94	95.48356	94.9743	83.64001	81.95908	83.0063
Dec 2018	75	90.71832	92.51853	80.44312	63.9463	71.95671

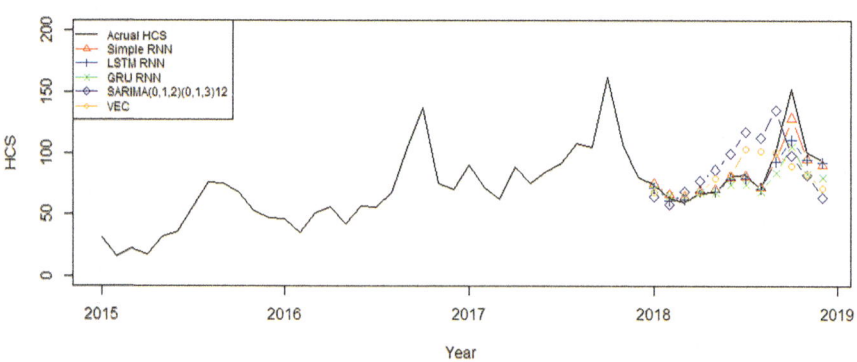

Fig. 5.13 HCS time series forecasts for 2018 using different RNNs and SARIMA model

5.8 Summary

This chapter presented the application of state-of-the-art recurrent neural networks (RNNs) for forecasting construction time series. It described how to prepare and preprocess time series data (i.e., lag transformation, scaling, and splitting) to use RNNs. Three different RNNs, including simple, LSTM, and GRU networks, were introduced, their structures explained, and their differences highlighted. The application of three different RNNs to forecast HCS time series for 12 months was illustrated. The prediction performances of the networks were evaluated and compared. The results showed that recurrent neural networks, especially LSTM and GRU networks, provide higher accuracies in forecasting the long-term variations of HCS than classical time series models, such as ARIMA, vector autoregressive, and vector error correction models. The trained networks had simple structures; hyperparameter tuning was not implemented to find the best hyperparameters for improving the performance of the networks. Therefore, sophisticated networks with hyperparameter tuning may offer much better forecasting performance.

5.9 Exercise Problems

Use the time series provided in the appendices to answer the problems.

1. Use the long short-term memory RNN to forecast the California Construction Cost Index (CCCI) for 12 months. Compare the results with the univariate AR forecasts presented in Chap. 1.
2. USE the long short-term memory and gated recurrent unit RNNs to forecast the National Highway Construction Cost Index (NHCCI). Compare the results with the ARIMA forecasts generated in Chap. 1.
3. Try to change the model specification (e.g., number of hidden layers, activation functions,…) for LSTM and GRU models to improve the model accuracy for HCS forecasts.

References

Abedinianjerabi, B., Shahandashti, S. M., Ahmadi, N., & Ashuri, B. (2017). Empirical investigation of temporal association between architecture billings index and construction spending using time-series methods. *Journal of Construction Engineering and Management, 143*(10), 04017080.

Bontempi, G., Taieb, S. B., & Le Borgne, Y. A. (2012). Machine learning strategies for time series forecasting. In *European business intelligence summer school* (pp. 62–77). Springer.

Chung, J., Gulcehre, C., Cho, K., & Bengio, Y. (2014). Empirical evaluation of gated recurrent neural networks on sequence modeling. *arXiv preprint arXiv:1412.3555*.

Dematos, G., Boyd, M. S., Kermanshahi, B., Kohzadi, N., & Kaastra, I. (1996). Feedforward versus recurrent neural networks for forecasting monthly Japanese yen exchange rates. *Financial Engineering and the Japanese Markets, 3*(1), 59–75.

Goodfellow, I., Bengio, Y., & Courville, A. (2016). *Deep learning*. MIT Press.

Graves, A. (2012). Supervised sequence labelling. In *Supervised sequence labelling with recurrent neural networks* (pp. 5–13). Springer.

Haykin, S. (2010). *Neural networks and learning machines* (3rd ed.). Pearson Education India.

Hemdan, E. E. D., Shouman, M. A., & Karar, M. E. (2020). Covidx-net: A framework of deep learning classifiers to diagnose covid-19 in x-ray images. *arXiv preprint arXiv:2003.11055*.

Joseph, V. R. (2022). Optimal ratio for data splitting. *Statistical Analysis and Data Mining: The ASA Data Science Journal, 15*, 531.

Mitchell, T. M. (1997). *Machine learning*. McGraw Hill.

Tealab, A. (2018). Time series forecasting using artificial neural networks methodologies: A systematic review. *Future Computing and Informatics Journal, 3*(2), 334–340.

Zhao, R., Wang, D., Yan, R., Mao, K., Shen, F., & Wang, J. (2017). Machine health monitoring using local feature-based gated recurrent unit networks. *IEEE Transactions on Industrial Electronics, 65*(2), 1539–1548.

Chapter 6
Investment Valuation of Construction Projects Under Uncertainty

Abstract Although many factors, including noneconomic barriers (e.g., regulatory and environmental factors), influence decision-making about construction projects, when it's time to invest, many construction project owners should allocate their limited financial resources to projects with the highest returns on investment. However, the investment valuation of construction projects is subject to significant uncertainties, such as substantial construction cost variations that make decision-making difficult. This chapter presents several investment valuation methods, such as a stochastic life-cycle cost analysis technique and a real options analysis method, to evaluate investments in construction projects under uncertainties. The stochastic life-cycle cost analysis captures the volatility of the input variables in investment valuation based on their historical values, propagates them through the life-cycle cost analysis method, and determines the probability distribution of the life-cycle cost. Real options analysis evaluates real (nonfinancial) investments under uncertainty with elements for strategic management flexibility and delayed investment. Various examples of construction investment valuations, along with the R codes, are presented in this chapter to enhance the learning experience. These resources can be extended for the assessment of other construction investment projects.

Keywords Construction investment valuation under uncertainty · Real options analysis · Stochastic life-cycle cost analysis · Binomial decision tree · Monte Carlo simulation · Investment decision-making under uncertainty

6.1 Stochastic Life-Cycle Cost Analysis of Construction Projects

Stochastic life-cycle cost analysis captures the volatility of the input variables in investment valuation based on their historical values, propagates them through the life-cycle cost analysis method, and determines the probability distribution of life-cycle cost (Fig. 6.1). Unlike the sensitivity analysis that provides only a range of

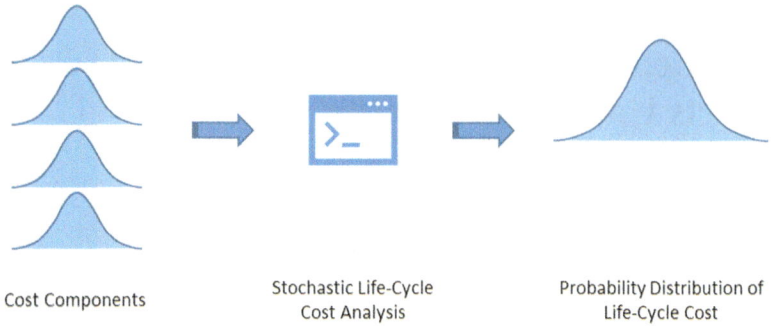

Fig. 6.1 Stochastic life-cycle cost analysis approach

outcomes (LCC), stochastic life-cycle cost analysis characterizes the probability distribution of the outcomes.

The veracity of the stochastic life-cycle cost analysis approach depends on selecting suitable probability distributions to represent the input variables. Probability distributions for input variables are often created using historical data. Previous research has identified many candidate probability distributions for input variables, such as triangular, normal, lognormal, Pareto, uniform, Weibull, and exponential distributions. In the case of abundant historical data on a cost component or an input variable, the parameters of a probability distribution can be estimated. Statistical goodness-of-fit tests, such as Kolmogorov-Smirnov test and the Cramér-von Mises criterion test (Darling, 1957; Arnold & Emerson, 2011), can be used to measure how good the distribution represents the data. In the case of limited available data, triangular distribution is often recommended to represent the underlying uncertain variables with the minimum, maximum, and most likely values that are set using decision-makers' judgment (Walls III & Smith, 1998; Campbell & Brown, 2003; Gransberg & Kelly, 2008; Burak Evrenosoglu, 2010; Pittenger et al., 2012). Monte Carlo simulation is often used to sample data from the input variables, propagate them into the life-cycle cost analysis method, and develop a probability distribution representing the life-cycle cost. Results can also be presented as empirical probability distributions or cumulative probability distributions; the latter is more intuitive for nonexperts. R Code 6.1 performs a Monte Carlo simulation using R software.

R Code 6.1
Monte Carlo simulation to calculate life-cycle cost of a construction project

```
LCCA_MC(comp1 = NA, comp2 = NA, comp3 = NA, comp4 = NA, comp5 = NA, recurring_comp = NA, r, n, n_loop)
```

6.1 Stochastic Life-Cycle Cost Analysis of Construction Projects

LCCA_MC is a function for calculating the present value of a project life-cycle cost using Monte Carlo simulation. The function is devised to incorporate up to five stochastic components (comp1 to comp5), recurring costs/benefits (recurring_comp) that happen every year, discount rate (r), service life, the project horizon in years (n), and the number of iterations (n_loop).

R Code 6.2 shows an example of implementing this code. First, the model components should be defined. In this example, component 1 is considered to have a triangular distribution with the minimum, maximum, and most likely values of $a = 10$, $b = 200$, and $c = 180$, respectively. Component 2 has a normal distribution with mean = 10 and sd = 10. Component 3 has a lognormal distribution with meanlog = 4 and sdlog = 0.5. Component 4 has a normal distribution with mean = 100 and sd = 20. And finally, component 5 has a uniform distribution with min = 100 and max = 300. This project also includes a fixed cost component that recurs annually with a value of 100. The discount rate r is set to be 5%, and the investment horizon is assumed to be 50 years. The model simulates the present value of the project life-cycle cost for 10,000 iterations.

R Code 6.2
Example of stochastic life-cycle cost analysis code

```
##General inputs
set.seed(42)    #This line of code is to fix the random number draw for the sake of teaching. You
may remove this line if you want to have slightly different answers.
library(triangle) #This line of code will activate the triangle library required for generating
random numbers.
n_loop = 10000
n = 50
##Defining input cost components
comp1 = rtriangle(n_loop, a =10, b = 200, c = 180)
comp2 = rnorm(n_loop, mean=10, sd=10)
comp3 = rlnorm(n_loop, meanlog = 4, sdlog = 0.5)
comp4 = rnorm(n_loop, mean=100, sd=20)
comp5 = runif(n_loop, min = 100, max = 300)
recurring_comp = 100

## Monte Carlo simulation to optain the project LCC
Project = LCCA_MC(comp1 = comp1, comp2 = comp2, comp3 = comp3, comp4 = comp4, comp5 = comp5,
r=0.05, n_loop=n_loop, n=n, recurring_comp = recurring_comp)

## Storing the simulation output as a .CSV file
write.csv(Project, " Project.csv")
```

Results

```
[1] 2280.482 2366.998 2363.614 2390.932 2489.677 2570.054 2579.930 2245.526
2411.516 2252.599 2249.461
[12] 2268.799 2442.332 2268.361 2228.069 2455.367 2449.676 2310.954 2296.131
2235.194 2414.473 2249.972
[23] 2257.976 2443.955 2358.174 2372.131 2294.982 2349.757 2483.431 2393.431
2206.297 2425.389 2333.409
[34] 2410.712 2301.647 2484.364 2221.343 2318.343 2441.085 2333.281 2304.984
2411.070 2225.400 2337.442.
.
.
[991] 2325.746 2198.202 2311.322 2519.453 2102.247 2369.475 2316.747 2172.548
2462.230 2255.863[ reached getOption("max.print") -- omitted 9000 entries ] [ reached getOption("ma
x.print") -- omitted 9000 entries ]
```

The model provides many outputs, such as the empirical probability distribution and cumulative probability distribution of the estimated present values of the project life cycle based on random outputs. The estimated present values of the project life cycle are stored in the computer directory as a comma-separated values (CSV) file. Figure 6.2 illustrates the empirical probability distribution and cumulative probability distribution of the project life-cycle cost.

From the distributions presented in Fig. 6.2, one can estimate the probability of a range of outcomes. Often, analysts use cumulative probability distributions for their simplicity to show the results to planners (Fig. 6.2b). It is straightforward to determine the probability of the life-cycle cost exceeding a certain threshold from the figure. Following the arrows, we can deduce that there is a 20% chance that the net present value of the project exceeds 2400. The model also provides the histogram of input components (Fig. 6.3). This information gives the decision-makers a better understanding of cost components in relation to the probability distribution of the net present values of life-cycle cost.

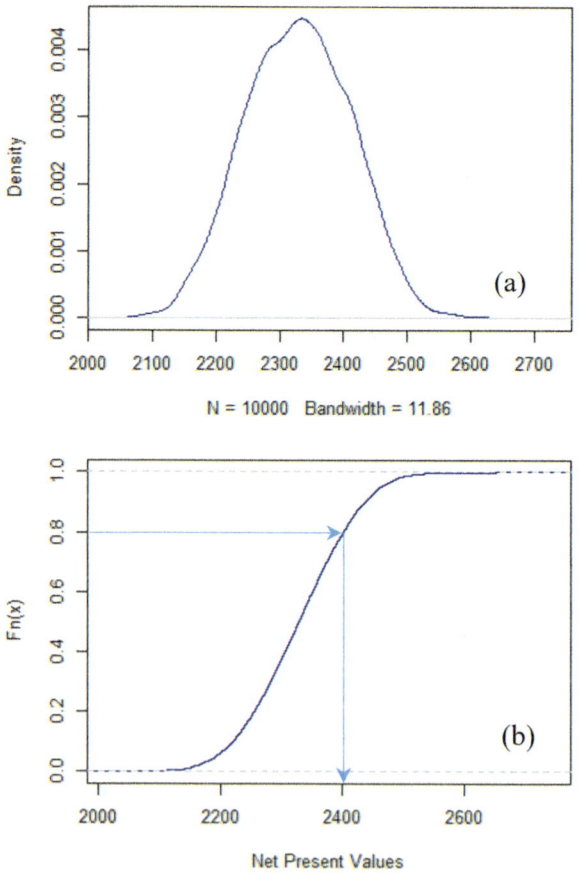

Fig. 6.2 Code outputs: (**a**) probability distribution of project life-cycle cost, (**b**) cumulative probability distribution of the project life-cycle cost

6.1 Stochastic Life-Cycle Cost Analysis of Construction Projects

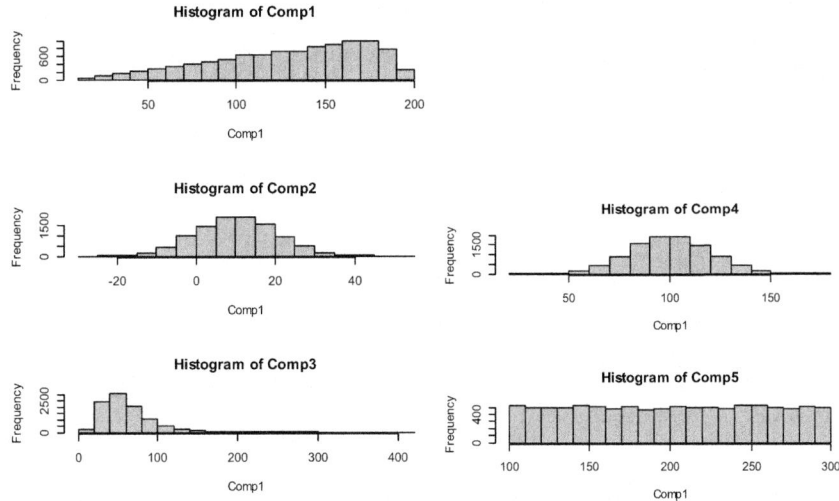

Fig. 6.3 Code outputs: histogram of input components of the life-cycle cost analysis

Although the presented Monte Carlo LCCA provides valuable graphs and good insight into the project life-cycle cost, the results of the Monte Carlo simulation can be further characterized (Jacquin & Shamseldin, 2007; Raychaudhuri, 2008; Thomopoulos, 2012) to help decision-making in the construction industry, where the investment values are often high and decisions are irreversible. A probability distribution function can be identified that best fits the empirical probability distribution resulting from the Monte Carlo LCCA. Parameters of the best-fitted distribution function (e.g., maximum, minimum, and average) are extremely valuable and often the only information one needs to make informed decisions. For instance, if the LCC distribution is best explained with a normal distribution, the average and standard deviation of the LCC normal distribution are helpful information to make informed decisions. Likewise, if the best distribution is triangular, one can determine the minimum, maximum, and most likely values of the project's LCC. R Code 6.3 finds the best distribution fit to the LCC probability distribution and acquires more accurate data.

R Code 6.3
Finding the best probability distribution that fits the life-cycle cost distribution

```
library(fitdistrplus)
require(mc2d)
# Fit distribution to get the parameters.
# Fit a normal distribution.
fn <- fitdist(Project, "norm")
summary(fn)
# Fit a uniform distribution.
fu <- fitdist(Project, "unif")
summary(fu)
# Fit a triangular distribution
# For triangular distribution, provide an estimate of min, max, and mode. this estimation can be
done using the dataset or the graphs obtained from the Monte Carlo simulation.
ft <- fitdist(Project, "triang", method="mge", start = list(min=2050, max=2600, mode=2325), gof
='CvM')
summary(ft)

#Plot the fitted distributions.
par(mfrow = c(2, 2))
plot.legend <- c("normal","uniform", "triangular")
denscomp(list(fn,fu, ft), legendtext = plot.legend)
qqcomp(list(fn,fu, ft), legendtext = plot.legend)
cdfcomp(list(fn,fu, ft), legendtext = plot.legend)
ppcomp(list(fn,fu, ft), legendtext = plot.legend)

# Conduct diagnosis Tests to see which one is the best fit.
gofstat(fn,fitnames = "norm")
gofstat(fu,fitnames = "unif")
gofstat(ft,fitnames = "triangle")
```

Results

```
> summary(fn)
Fitting of the distribution ' norm ' by maximum likelihood
Parameters :
       estimate Std. Error
mean 2328.03752  0.8312299
sd     83.12297  0.5877683
Loglikelihood:  -58392.6   AIC:  116789.2   BIC:  116803.6
Correlation matrix:
     mean sd
mean    1  0
sd      0  1
> summary(fu)
Fitting of the distribution ' unif ' by maximum likelihood
Parameters :
    estimate Std. Error
min 2062.559         NA
max 2693.766         NA
Loglikelihood:  -64476.35   AIC:  128956.7   BIC:  128971.1
Correlation matrix:
[1] NA
> summary(ft)
Fitting of the distribution ' triang ' by maximum goodness-of-fit
Parameters :
       estimate
min  2127.398
max  2528.621
mode 2327.775
Loglikelihood:  -Inf    AIC:  Inf   BIC:  Inf
```

See Appendix N for detailed results.

Based on experience, most stochastic LCCA results are more likely to have a normal, uniform, or triangular distribution. Therefore, these probabilistic distributions are first fit to the calculated LCC values, and then the goodness-of-fit tests are used to identify which one is the best fit. Kolmogorov-Smirnov, Cramér-von Mises,

6.1 Stochastic Life-Cycle Cost Analysis of Construction Projects

and Anderson-Darling goodness-of-fit tests were included in R Code 6.3. The corresponding best-fitted distribution parameters are the information we are looking for. In this example, the normal distribution is the best fit for the empirical distribution of the LCC of the project. From the results, the most likely expected value of the project's LCC is around 2328, with a standard deviation of 83. Since the fundamentals behind the goodness-of-fit tests and distribution functions are out of the scope of this book, the detailed outputs of running R Code 6.3 are only provided in Appendix N for interested readers. R Code 6.3 also plots histogram and theoretical densities, Q-Q plot, empirical and theoretical cumulative distribution functions, and P-P plot. These plots are shown in Fig. 6.4. Figure 6.4 shows that the normal distribution is the best fit for the present value of the project life-cycle cost (red curves).

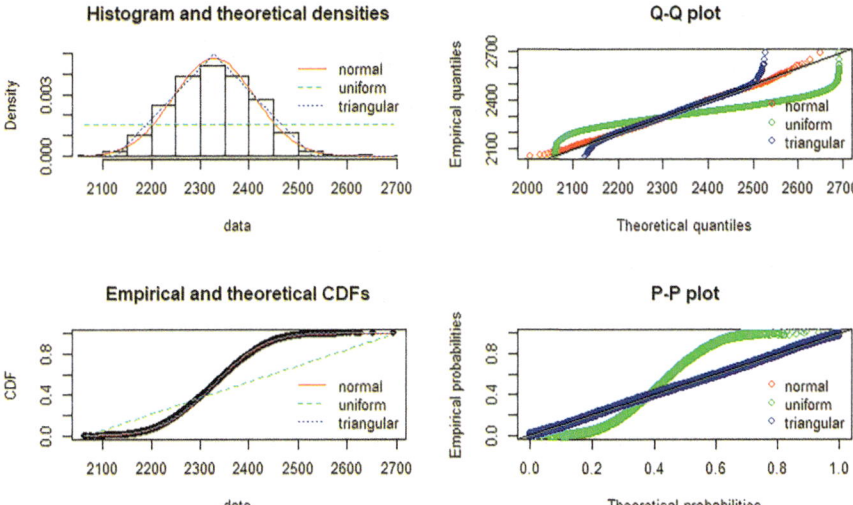

Fig. 6.4 Graphical results of goodness-of-fit tests

Example 6.1

A state department of transportation decides to build a new two-lane road between two villages 10 miles apart. Due to limited funding sources and the growing number of high-priority projects competing for funding, a detailed investment valuation that considers cost estimation risks is required. Based on the data provided in Table 6.1, perform a stochastic life-cycle cost analysis for the project. Assume a service life of 35 years and an annual discount rate of 2.5% for this project.

Solution R Codes 6.2 and 6.3 were updated with the information provided in Example 6.1, resulting in R Code 6.4. R Code 6.4 performs stochastic life-cycle cost analysis for the two-lane road construction project. The min, max, and mode values for fitting the triangular distribution are set with trial and error.

Figure 6.5 illustrates the results of the stochastic life-cycle cost analysis. This figure shows the empirical probability distribution and cumulative probability distribution of the project life-cycle cost.

Table 6.1 Input variables of the two-lane road construction project

Input variable	Distribution type	Description
Engineering design	Normal	Mean = $500 K, SD = $200 K
Right-of-way (ROW) preparation	Weibull	$20 million, shape factor = 0.05, scale number = 0.2
Pavement construction	Triangular	a = $80 million, b = $120 million, c = $130 million
Signage and pavement markings	Determinate	$5 million
Annual maintenance	Normal	Mean = $500 K, SD = $200 K

R Code 6.4
Stochastic life-cycle cycle cost analysis of the two-lane road construction project

```
library(triangle)
library(fitdistrplus)
require(mc2d)
##General inputs
set.seed(22)
n_loop = 1000
n = 35
comp1_A = rnorm(n_loop, mean=0.5, sd=0.2)
comp2_A = rweibull(n_loop, shape=0.5, scale=0.2)
comp3_A = rtriangle(n_loop, a =80, b = 130, c = 120)
comp4_A = rep(5, length.out=n_loop)
recurring_comp = rnorm(n, mean=0.5, sd=0.2)
Project_A = LCCA_MC(comp1 = comp1_A, comp2 = comp2_A, comp3 = comp3_A, comp4 = comp4_A, r=0.025, n_loop=n_loop, n=n, recurring_comp = recurring_comp)
# Fit distribution to get the parameters.
fn <- fitdist(Project_A, "norm")
summary(fn)
fu <- fitdist(Project_A, "unif")
summary(fu)
ft <- fitdist(Project_A, "triang", method="mge", start = list(min=93, max=153, mode=135), gof ='CvM')
summary(ft)
par(mfrow = c(2, 2))
plot.legend <- c("normal","uniform", "triangular")
denscomp(list(fn,fu, ft), legendtext = plot.legend)
qqcomp(list(fn,fu, ft), legendtext = plot.legend)
cdfcomp(list(fn,fu, ft), legendtext = plot.legend)
ppcomp(list(fn,fu, ft), legendtext = plot.legend)
# Conduct the Fit Test to see which one is the best fit.
gofstat(fn,fitnames = "norm")
gofstat(fu,fitnames = "unif")
gofstat(ft,fitnames = "triangle")
```

6.1 Stochastic Life-Cycle Cost Analysis of Construction Projects

Results

```
> summary(fn)
Fitting of the distribution ' norm ' by maximum likelihood
Parameters :
      estimate Std. Error
mean 127.26991 0.3413578
sd    10.79468 0.2413764
Loglikelihood:  -3797.992   AIC:  7599.984   BIC:  7609.8
> summary(fu)
Fitting of the distribution ' unif ' by maximum likelihood
Parameters :
    estimate Std. Error
min  97.50048         NA
max 147.17710         NA
Loglikelihood:  -3905.534   AIC:  7815.069   BIC:  7824.884
summary(ft)
Fitting of the distribution ' triang ' by maximum goodness-of-fit
Parameters :
     estimate
min   98.10512
max  147.10981
mode 136.87340
Loglikelihood:  -Inf   AIC:  Inf   BIC:  Inf
```

See Appendix N for detailed results.

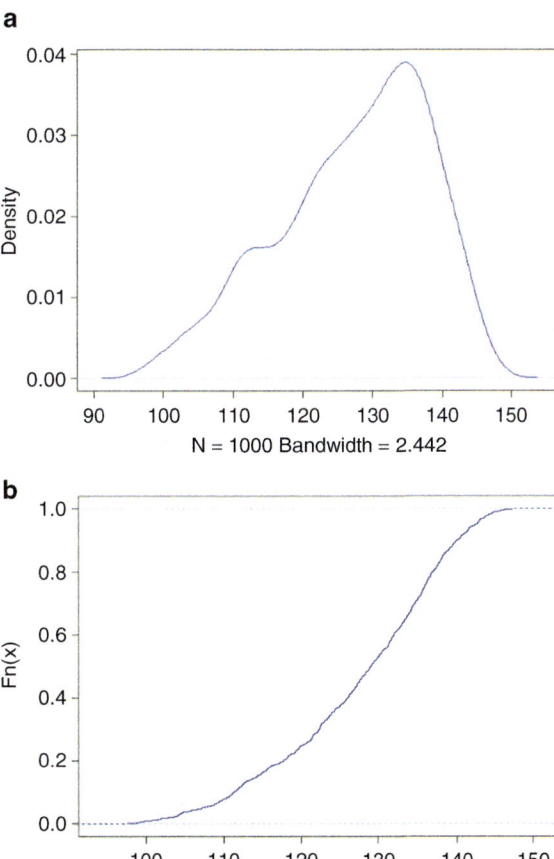

Fig. 6.5 (**a**) Probability distribution of the project life-cycle cost, (**b**) cumulative probability distribution of the project life-cycle cost

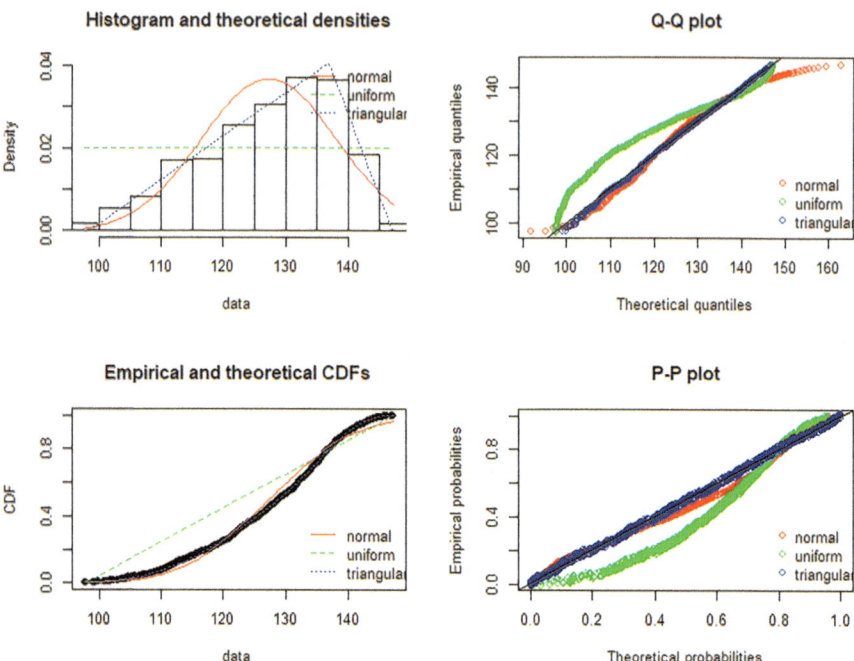

Fig. 6.6 Results of the stochastic LCCA of the two-lane road construction project

Figure 6.6 shows that the best fit to the project life-cycle cost distribution is a triangular distribution with min, max, and mode values of 98.10, 147.11, and 136.87, respectively. Please see Appendix N for the complete analysis results and code outputs. The results show that the project will likely cost around $136 million.

6.2 Life-Cycle Cost Comparison of Alternative Construction Projects

Stochastic life-cycle cost analyses of alternative construction projects provide essential information to make risk-based decisions considering the uncertain future. Figure 6.7 shows the probability distribution and cumulative probability distribution for two alternative construction projects. The figure shows that project A has a lower expected life-cycle cost compared to project B, which ostensibly makes project A more appealing. However, the high variation of life-cycle costs indicates that project A is riskier; it is more probable that project A becomes over budget or under budget (if the expected value is used as the benchmark). On the contrary, project B has a

Fig. 6.7 Risk profile of two alternative projects

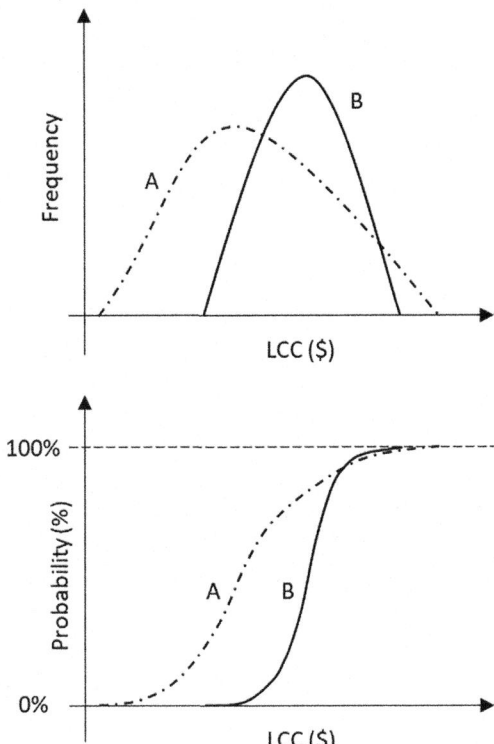

higher expected cost with lower-cost variations. A risk-averse decision-maker may select project B, regardless of its higher expected cost. The selection of an alternative project depends on the project decision-maker's risk appetite.

With minor modifications, the proposed approach and R codes can be used to conduct stochastic life-cycle benefit-cost analysis based on uncertain cost and benefit inputs represented by various probability distributions. It is recommended to use a negative sign for a cost component and a positive sign for a revenue component while building the cash flow for the project alternatives. Example 6.2 provides an illustrative case on how to perform a stochastic life-cycle benefit-cost analysis of two alternative projects to support risk-based decision-making.

Example 6.2

A housing construction company plans to invest in a 4-acre land in the middle of a fully developed area in a major city. The consulting firm came up with two competing ideas: (1) building an apartment complex with 250 units or (2) building 50 single-family rental townhomes. Based on the information provided in Tables 6.2

Table 6.2 Cost information on apartment complex scenario

Items	Cost and metrics	Distribution
Engineering and architectural design	$\mu = -\$1$ million SD = \$200 K	Normal
Overhead costs and permitting	−\$3 million	Fixed
Construction cost	$a = -\$80$ million $b = -\$35$ million $c = -\$50$ million	Triangular
Annual maintenance	$a = -\$3$ million $b = -\$1.5$ million $c = -\$2$ million	Triangular
Annual revenue	$\mu = \$6$ million SD = \$500 K	Normal

Table 6.3 Cost information on rental townhome scenario

Items	Cost and metrics	Distribution
Engineering and architectural design	$\mu = -\$0.7$ million SD = \$150 K	Normal
Overhead costs and permitting	−\$2.5 million	Fixed
Construction cost	$a = -\$35$ million $b = -\$10$ million $c = -\$15$ million	Triangular
Annual maintenance	$a = -\$1$ million $b = -\$0.3$ million $c = -\$0.5$ million	Triangular
Annual revenue	$\mu = \$2.6$ million SD = \$200 K	Normal

and 6.3, conduct a stochastic life-cycle benefit-cost analysis to conduct a risk-based comparison of the alternative project. Consider a service life of 30 years and a 4% interest rate.

Solution R Code 6.5 was used to conduct the stochastic life-cycle benefit-cost analysis for two alternative projects.

6.2 Life-Cycle Cost Comparison of Alternative Construction Projects

R Code 6.5
Stochastic life-cycle benefit-cost analysis of two alternative projects

```
library(triangle)
##General inputs
set.seed(22)
n_loop = 1000
n = 30

##Specific inputs for Project A (Apartment complex)
comp1_A = rnorm(n_loop, mean=-1, sd=0.2)
comp2_A = rep(-3, length.out=n_loop)
comp3_A = rtriangle(n_loop, a =-80, b = -35, c = -50)
recurring_comp_A = rnorm(n, mean=6, sd=0.5) + rtriangle(n, a =-3, b = -1, c = -2)
Project_A = EUAC_MC(comp1 = comp1_A, comp2 = comp2_A, comp3 = comp3_A,r=0.04, n_loop=n_loop, n=n,
recurring_comp = recurring_comp_A)

##Specific inputs for Project B (Rental townhome)
comp1_B = rnorm(n_loop, mean=-0.7, sd=0.15)
comp2_B = rep(-2.5, length.out=n_loop)
comp3_B = rtriangle(n_loop, a =-35, b = -10, c = -15)
recurring_comp_B = rnorm(n, mean=2.8, sd=0.2) + rtriangle(n, a =-1, b = -0.3, c = -0.5)
Project_B = EUAC_MC(comp1 = comp1_B, comp2 = comp2_B, comp3 = comp3_B, r=0.04, n_loop=n_loop, n=n,
recurring_comp = recurring_comp_B)

##Density Plot
plot(density(Project_A),col="blue",main="Kernel Density of Equivalent Annual Cost - Project A
versus Project B",xlab="Net Present Values",ylim=c(0, 1.5))
lines(density(Project_B),col="red")
legend("topright",
legend = c("Project A", "Project B"),
col = c("blue", "red"),
pch = c(16,16),
bty = "n",
pt.cex = 1,
cex = 1.2,
text.col = "black",
horiz = F)


##Cumulative Density Plot
plot(ecdf(Project_A),main="Cumulative Density Function of Equivalent Annual Cost - Project A versus
Project B",
xlab="Net Present Values",col="blue")
lines(ecdf(Project_B),col="red")
legend("topright",
legend = c("Project A", "Project B"),
col = c("blue", "red"),
pch = c(16,16),
bty = "n",
pt.cex = 1,
cex = 1.2,
text.col = "black",
horiz = F)
Project_Eval = cbind(Project_A, Project_B)
write.csv(Project_Eval, "Project_Eval.csv")#write.csv(Project_Eval, "Project_Eval.csv")
```

Figure 6.8 shows the resulting risk profiles of two alternative projects. Project A (apartment complex) has a lower, most likely life-cycle cost compared to project B, while the LCC of project B is less uncertain, and one can be more definite about the

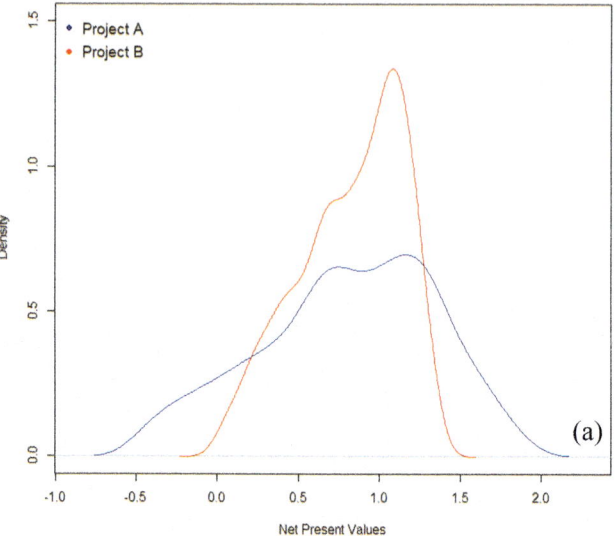

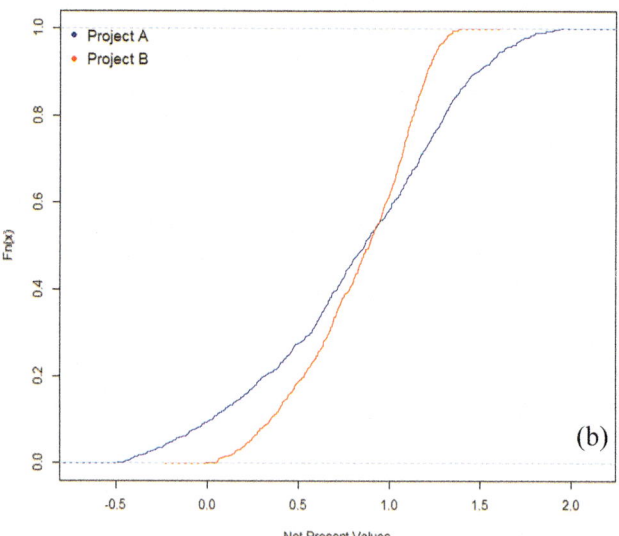

Fig. 6.8 (**a**) Risk profile of two alternative residential projects, (**b**) cumulative distribution of two alternative residential projects

6.2 Life-Cycle Cost Comparison of Alternative Construction Projects

values that were obtained from the analysis. A risk-averse builder may select project B since they might find planning for a high cost with a smaller range of possible variations more suitable.

Let's perform a goodness-of-fit test on the project LCCs. Run the R code provided in R Code 6.6 to see the results. This section highlights some results; a full detailed version of the results is presented in Appendix N. The goodness-of-fit test suggests that both project LCCs are better explained with a triangular distribution.

R Code 6.6
Goodness-of-fit test for two alternative projects

```
library(fitdistrplus)
descdist(Project_A)
descdist(Project_B)
#fw_A <- fitdist(Project_A, "weibull")
#fg_A <- fitdist(Project_A, "gamma")
#fln_A <- fitdist(Project_A, "lnorm")
fn_A <- fitdist(Project_A, "norm")
summary(fn_A)
fu_A <- fitdist(Project_A, "unif")
summary(fu_A)
ft_A <- fitdist(Project_A, "triang", method="mge", start = list(min=-0.5, max=1.5, mode=0.6), gof = "CvM")
summary(ft_A)

fn_B <- fitdist(Project_B, "norm")
summary(fn_B)
fu_B <- fitdist(Project_B, "unif")
summary(fu_B)
ft_B <- fitdist(Project_B, "triang", method="mge", start = list(min=-0.5, max=1.5, mode=0.6) , gof = "CvM")
summary(ft_B)

par(mfrow = c(2, 2))
plot.legend <- c("normal","uniform", "triangular")
denscomp(list(fn_A,fu_A, ft_A), legendtext = plot.legend)
qqcomp(list(fn_A,fu_A, ft_A), legendtext = plot.legend)
cdfcomp(list(fn_A,fu_A, ft_A), legendtext = plot.legend)
ppcomp(list(fn_A,fu_A, ft_A), legendtext = plot.legend)
gofstat(fn_A,fitnames = "norm")
gofstat(fu_A,fitnames = "unif")
gofstat(ft_A,fitnames = "triangle")

par(mfrow = c(2, 2))
plot.legend <- c("normal","uniform", "triangular")
denscomp(list(fn_B,fu_B, ft_B), legendtext = plot.legend)
qqcomp(list(fn_B,fu_B, ft_B), legendtext = plot.legend)
cdfcomp(list(fn_B,fu_B, ft_B), legendtext = plot.legend)
ppcomp(list(fn_B,fu_B, ft_B), legendtext = plot.legend)
gofstat(fn_B,fitnames = "norm")
gofstat(fu_B,fitnames = "unif")
gofstat(ft_B,fitnames = "triangle")
```

Results

```
#Goodness-of-fit test for Project A
Goodness-of-fit criteria
unif
Akaike's Information Criterion 1788.948
Bayesian Information Criterion 1798.763
> gofstat(ft_A,fitnames = "triangle")
Goodness-of-fit statistics
triangle
Kolmogorov-Smirnov statistic 0.02200626
Cramer-von Mises statistic 0.05253836
Anderson-Darling statistic Inf

Goodness-of-fit criteria
triangle
Akaike's Information Criterion Inf
Bayesian Information Criterion Inf

#Goodness-of-fit test for Project B

Goodness-of-fit criteria
unif
Akaike's Information Criterion 660.1680
Bayesian Information Criterion 669.9836
> gofstat(ft_B,fitnames = "triangle")
Goodness-of-fit statistics
triangle
Kolmogorov-Smirnov statistic 0.01685045
Cramer-von Mises statistic 0.04643023
Anderson-Darling statistic Inf
Goodness-of-fit criteria
triangle
Akaike's Information Criterion Inf
Bayesian Information Criterion Inf
```

See Appendix N for detailed results.

6.3 Real Options Analysis of Construction Projects

Conventional investment valuation methods (e.g., net present value analysis) assume that all construction investment decisions are made at once and are irrevocable. These assumptions are not consistent with real-world construction investment decision-making (especially innovative and risky projects), in which an investor can delay the construction investment decision. Real options analysis evaluates real (nonfinancial) investments with elements for strategic management flexibility and delayed investment (Dixit & Pindyck, 1994). In construction analytics, a real option is a right but not the obligation to make a construction business decision. Real options analysis is a valuable method for investment valuation of construction projects under uncertainty. Over the past couple of decades, real options analysis has been recognized as a promising approach for the investment valuation of many innovative construction and infrastructure projects under various uncertainties. For example, while Zahed et al. (2020) used real options to evaluate underground freight transportation systems under construction cost and freight rate uncertainties, Kashani et al. (2015) used it to identify the best time to implement renewable energy systems in buildings. The literature includes many real options models developed based on various mathematical formulations. The binomial tree model is one of the most explicable yet rigorous approaches.

6.4 Binomial Tree Model

The binomial option pricing model is a well-established method for pricing both financial and real options, initially proposed by Cox et al. (1979). The binomial tree model presented in this chapter is based on Guthrie's (2009) representation of the binomial option pricing model. In this approach, a state variable (uncertain variable), denoted by X, is selected. The state variable represents the uncertain underlying asset. This variable is unaffected by a decision-maker's actions and describes the primary source of risk that affects the investment valuation of the project. The binomial option pricing model assumes that an option's life, T, can be divided into n equal discrete periods, Δt, and the asset value will increase or decrease according to a binomial process. The underlying asset (state variable) may either increase or decrease over each discrete period, t, by a move up factor, U ($U > 1$) with probability θ_u, or a move down factor, D ($D < 1$) with probability θ_d (which $\theta_d = 1 - \theta_u$). The two states that the underlying asset can take after a period are shown in Fig. 6.9.

The growth factors U and D represent how much the state variable has increased or decreased after a period. Considering the stochastic behavior of the state variable, the growth factors are characterized as exponential functions of volatility (σ) and time (Δt) (Hull, 2017; Cox et al., 1979):

$$U = e^{\sigma\sqrt{\Delta t}} \quad \text{and} \quad D = 1/U = e^{-\sigma\sqrt{\Delta t}} \tag{6.1}$$

Starting with the asset's base value and ending with the option's expiration (i.e., the time frame within which the decision should be made), the asset extends into a binomial tree that indicates a sequence of up or down movements. Figure 6.10 illustrates such a binomial tree. Based on the potential values of the state variable at each time step, several project cash flows can be estimated. If any node at time n, $n = 1, 2, \ldots, T$, can be reached with i number of down moves, $0 \leq i \leq n$, and $n - i$ number of up moves, $0 \leq n - i \leq n$, the value of X at that node would be calculated as $X(i,n) = X_0 U^{n-i} D^i$. Based on the binomial probability mass function, the probability of $C(i,n)$ to occur is $P[C(i,n)] = \binom{n}{i} \theta_u^{n-i} (1-\theta_u)^i$.

Fig. 6.9 Asset price binomial tree over one period

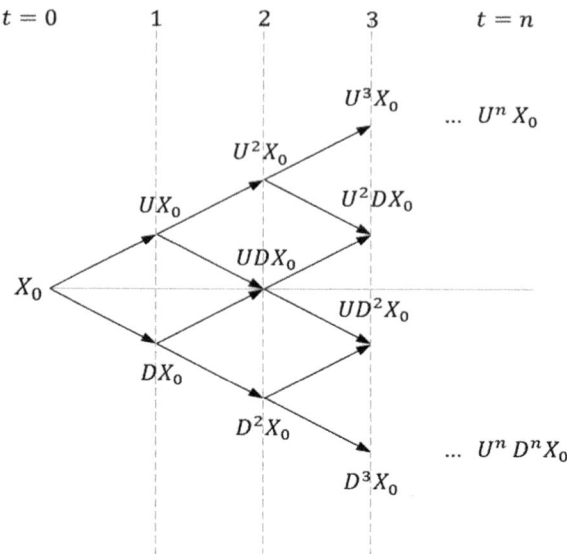

Fig. 6.10 The binomial tree of the state variable

Regardless of how the decision-maker behaves toward risk, the main objective is to estimate the market value of the project by estimating the value of the future possible cash flows. As a functional method, arbitrage-free asset pricing is frequently utilized to determine the market value of future cash flows and ultimately price the value of the available real options. The notion is to assume the existence of a portfolio of traded assets, including a risk-free bond and a risky asset, which generates a cash flow stream approximately equal to the one being valued (Guthrie, 2009). The value of this replicating portfolio is anticipated as the expected market value of estimates. Suppose a project generates cash flows of Y_u in the up state and Y_d in the down state after a period. The expected value of these future cash flows at the current time is:

$$V = \frac{\pi_u Y_u + \pi_d Y_d}{R_f}, \tag{6.2}$$

where π_u and π_d are the risk-neutral probabilities of up and down moves, while R_f is the periodic growth factor equal to $1 + r_f$ and r_f is the risk-free interest rate. Using the capital asset pricing model (CAPM), the risk-neutral probabilities can be presented as:

$$\pi_u = \frac{K - D}{U - D} \quad \text{and} \quad \pi_d = \frac{U - K}{U - D}, \tag{6.3}$$

6.4 Binomial Tree Model

where K is the risk-adjusted growth factor. Various estimations of K have been presented in the literature based on different underlying assumptions. With the assumption of a risk-neutral environment and no arbitrage, Cox et al. (1979), Copeland and Antikarov (2001), Björk (2009), and Hull (2017) showed that the tracking portfolio has risk-free future payoffs, which can be discounted using a risk-free interest rate. They supposed that the best unbiased estimate of the market is in fact the present value of the project itself (without flexibility). They viewed K as a growth factor that is broadly similar to the one-period risk-free rate of return and can be valued as $K = (1 + r_f \Delta t)$ or $K = e^{r_f(\Delta t)}$. Although this assumption could be found extreme, Copeland and Antikarov (2001) believe that "it is no stronger than those used to estimate the project NPV in the first place." Alternatively, Hull (2017) and Guthrie (2009) presented another perspective, as the volatility of a project could be assumed to be almost the same as the volatility of its underlying risky asset. They defined K as a risk-adjusted growth factor that includes the market price of the risk. Further, they estimated the martingale measures and the market value of the option by performing risk-neutral pricing. Hull (2017) presented the market price of the risk as $\lambda = (\mu - r)/\sigma$. Rendleman Jr (1999) and Guthrie (2009) adjusted the model for the market price of the risk by applying the capital asset pricing model (CAPM). They showed that K could be estimated as the difference between the CAPM risk premium $\left(E\left[\tilde{R}_m\right] - R_x\right)\beta_x$ and the expected growth factor of the state variable $E\left[\tilde{R}_x\right]$ (Guthrie, 2009):

$$K = E\left[\tilde{R}_x\right] - \left(E\left[\tilde{R}_m\right] - R_x\right)\beta_x. \tag{6.4}$$

Implementing a similar concept, the value of a cash flow generated by the state variable at a future time can be determined using the backward induction method. If we have the possible values of the cash flow stream at a selected time y_t, the expected values at a time step earlier can be calculated using Eq. (6.2). Using Eq. (6.5), moving backward to the current date, the value of the cash flow stream at each time step can be determined.

$$V(i,n) = \frac{\pi_u(i,n)V(i,n+1) + \pi_d(i,n)V(i+1,n+1)}{R_f}. \tag{6.5}$$

Similarly, the value of a series of possible future cash flows is calculated using Eq. (6.6) and backward induction

$$V(i,n) = Y(i,n) + \frac{\pi_u(i,n)V(i,n+1) + \pi_d(i,n)V(i+1,n+1)}{R_f}. \tag{6.6}$$

Having the expected value of the project cash flows at each possible condition in the future would help managers foresee the possible outcomes and prepare them for making decisions. A dynamic programming technique can be employed to maximize the outcomes of an investment option. Assume a decision-maker wants to evaluate an investment project. At each node, the decision-maker has the option of investing in the project or standing by until the following period to collect more market information. The decision-maker can calculate and compare the value of both actions to make the most optimal decision at each node. This technique is recognized as Bellman's "principle of optimality," where a complicated multi-period optimization problem is divided into a series of simpler problems. All the pertinent information about the future is summarized at each node by bringing back (backward induction) the possible following cash flows. Using the information, the decision-maker can compare the value of investing or not investing in the project and select an option with the highest outcome. To simply put, the market value of the investment project at each node is:

$$V'(i,n) = \max\{V_{\text{not invest}}(i,n), V_{\text{invest}}(i,n)\}. \tag{6.7}$$

Depending on the investment project, a suitable valuation equation should be designed and used to identify the best options.

Example 6.3
A residential development company has a 5-year window to invest in an apartment complex project with a construction cost of $50 million. Based on company estimates, the construction cost would increase at a rate of $\alpha = 4\%$ per year. It is assumed that the construction process would take 1 year, and the current value of the completed complex is $50 million. Suppose the firm's interest rate is 5%, the volatility of the real estate market is 10%, and the risk-adjusted growth factor K is 1.01. Considering these conditions, when would be the best time for the company to invest in the project?

Solution The first step is to determine the model input variables using the provided information. These values are as follows:

$X_0 = \$50$ million	$\sigma = 0.1$
$I = \$50$ million	$U = e^{\sigma\sqrt{\Delta t}} = 1.105$
$\Delta t = 1$ year	
$r = 0.05$	$D = 1/U = 0.905$
$R_f = 1 + r = 1.05$	$\pi_u = \dfrac{K - D}{U - D} = 0.525$
$K = 1.01$	
$\alpha = 4\%$	$\pi_d = \dfrac{U - K}{U - D} = 1 - \pi_u = 0.475$

6.4 Binomial Tree Model

The next step is to create the binomial tree of the facility value based on the market conditions. The value of the finished complex apartment at each node of the binomial tree can be obtained from $X(i,n) = X_0 U^{n-i} D^i$. To find the value of the option to invest in the project, the company should decide between investing in the project and waiting for another period to gain more information on the market. If the company decides to invest in the project at the node (i,n), a construction cost of $I = \$50$ million is charged immediately, and the project can be sold after a year which the construction is finished. Technically, the value of the project, if they decide to invest, would be:

$$V_I(i,n) = -I + \frac{\pi_u(i,n)X(i,n+1) + \pi_d(i,n)X(i+1,n+1)}{R_f}. \tag{6.8}$$

Otherwise, if the company decides to wait for another period, the value of the project would be the expected value of investing in the project a year later, whether the market goes up or down. In other words, the value of waiting would be:

$$V_W(i,n) = \frac{\pi_u(i,n)V_I(i,n+1) + \pi_d(i,n)V_I(i+1,n+1)}{R_f}. \tag{6.9}$$

Therefore, the value of the option to invest in this project can be obtained from:

$$V'(i,n) = \max \left\{ \begin{array}{l} -I + \dfrac{\pi_u(i,n)X(i,n+1) + \pi_d(i,n)X(i+1,n+1)}{R_f}, \\ \dfrac{\pi_u(i,n)V_I(i,n+1) + \pi_d(i,n)V_I(i+1,n+1)}{R_f} \end{array} \right\} \tag{6.10}$$

R Code 6.7 was used to create the binomial tree model and find the best strategy to invest in the project.

Figure 6.11 shows the binomial tree for the cost of the completed complex (state variable). It is shown that the cost of implementing the project would go up or down

---

R Code 6.7

Optimal time to invest using the binomial tree model

```
# I is defined as a vector of investment cost that grows with α=4%. Due to Time=5, the size of this
vector should be set equal to 5.
BinomialTree(S=50, I=cumprod(c(50, rep(1+0.04, 5))), Time=5, r=0.05, sigma=0.1, dt=1, k =1.01,
imm =FALSE)
```

Results

```
[1] "Model Parameters:"
S      Time   Rf     sigma   n    k      Up     Down    Pi_Up   Pi_Down
50     5      1.05   0.1     5    1.01   1.105  0.905   0.525   0.475
```

[1] "Binomial Tree:"

	0	1	2	3	4	5	6
0	50	55.26	61.07	67.49	74.59	82.44	91.11
1	0	45.24	50	55.26	61.07	67.49	74.59
2	0	0	40.94	45.24	50	55.26	61.07
3	0	0	0	37.04	40.94	45.24	50
4	0	0	0	0	33.52	37.04	40.94
5	0	0	0	0	0	30.33	33.52
6	0	0	0	0	0	0	27.44

[1] "Cashflow:"

	0	1	2	3	4	5	6
0	-1.90	3.15	8.74	14.92	21.75	29.30	0.00
1	0	-8.48	-3.90	1.15	6.74	12.92	0.00
2	0	0	-14.70	-10.56	-5.98	-0.93	0.00
3	0	0	0	-20.61	-16.87	-12.72	0.00
4	0	0	0	0	-26.25	-22.86	0.00
5	0	0	0	0	0	-31.66	0.00
6	0	0	0	0	0	0	0.00

[1] "Option Value:"

	0	1	2	3	4	5	6
0	3.01	5.25	8.99	14.92	21.75	29.30	0.00
1	0	0.84	1.69	3.37	6.74	12.92	0.00
2	0	0	0.00	0.00	0.00	0.00	0.00
3	0	0	0	0.00	0.00	0.00	0.00
4	0	0	0	0	0.00	0.00	0.00
5	0	0	0	0	0	0.00	0.00
6	0	0	0	0	0	0	0.00

[1] "Decision Tree:"

	Year 0	Year 1	Year 2	Year 3	Year 4	Year 5
0	Wait	Wait	Wait	Invest	Invest	Invest
1	–	Wait	Wait	Wait	Invest	Invest
2	–	–	Wait	Wait	Wait	Wait
3	–	–	–	Wait	Wait	Wait
4	–	–	–	–	Wait	Wait
5	–	–	–	–	–	Wait

every year with a magnitude of U and D, respectively. Based on this binomial tree, the project cash flow and the value of the investment option can be derived. The Results section of R Code 6.7 summarizes the input variables and presents the four binomial trees, including the binomial tree of the state variable, the binomial tree of the project cash flow, the binomial tree of investment option values, and, finally, the decision tree. The decision tree indicates that this project is not economical in the first 2 years. It suggests that a manager should wait for at least 3 years and evaluate the market to decide whether invest in the project or not. Let's assume that the market goes up for 3 consecutive years (node (3,0)); based on our evaluation, this project would become economical to implement. We can also anticipate that waiting for another year would lead us to a situation that still the project is economical, whether the market goes down in year 4 (node (4,1)). The decision tree provides valuable information for managers, giving them more flexibility in making decisions based on market volatility.

Further, Monte Carlo simulation can be used to determine the likelihood of undertaking the investment decision. In this approach, the model takes numerous random paths to reach a node in the binomial tree, conducts all the calculations, and

6.4 Binomial Tree Model

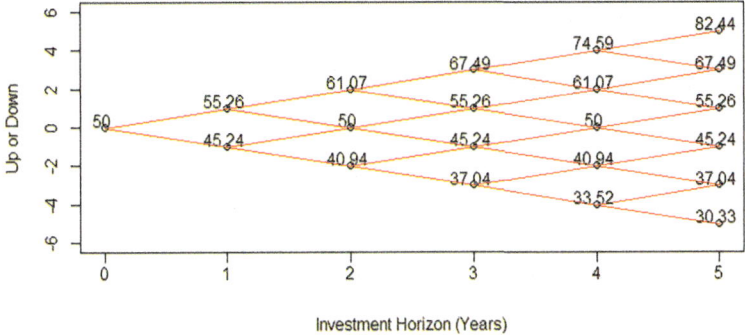

Fig. 6.11 The binomial tree of the state variable for Example 6.3

finally determines the optimal investment decision. Using all these data, we can determine the likelihood of reaching conditions (nodes) that executing the investment option is economical. Example 6.4 presents the use of this technique in more detail.

Solution R Code 6.8 was used to create the binomial tree model and find the best strategy to invest in the project using Monte Carlo simulation.

Example 6.4
Determine the likelihood of investment in the project described in Example 6.3 using the Monte Carlo simulation method.

Since the input variables are not changed, the results shown in R Code 6.8 is similar to what is presented in the Results section of R Code 6.7. In addition, R Code 6.8

R Code 6.8
Optimal time to invest using binomial tree model

```
library(plyr)
library(ggplot2)
BinomialTree_MC(S=50, I=cumprod(c(50, rep(1+0.04, 5))), Time=5, r=.05, sigma=0.1, dt=1, imm = FALSE,
k=1.01, MC_loops = 1000)
```

Results

```
[1] "Model Parameters:"
S     Time   Rf     sigma   n    k      Up     Down    Pi_Up   Pi_Down
50    5      1.05   0.1     5    1.01   1.105  0.905   0.525   0.475
[1] "Binomial Tree:"
       0      1      2       3       4       5
0      50     55.26  61.07   67.49   74.59   82.44
1      0      45.24  50      55.26   61.07   67.49
2      0      0      40.94   45.24   50      55.26
3      0      0      0       37.04   40.94   45.24
4      0      0      0       0       33.52   37.04
5      0      0      0       0       0       30.33
6      0      0      0       0       0       0
[1] "Cashflow:"
       0      1      2       3       4       5       6
0      -1.90  3.15   8.74    14.92   21.75   29.30   0.00
1      0      -8.48  -3.90   1.15    6.74    12.92   0.00
2      0      0      -14.70  -10.56  -5.98   -0.93   0.00
3      0      0      0       -20.61  -16.87  -12.72  0.00
4      0      0      0       0       -26.25  -22.86  0.00
5      0      0      0       0       0       -31.66  0.00
6      0      0      0       0       0       0       0.00
[1] "Option Value:"
       0      1      2       3       4       5       6
0      3.01   5.25   8.99    14.92   21.75   29.30   0.00
1      0      0.84   1.69    3.37    6.74    12.92   0.00
2      0      0      0.00    0.00    0.00    0.00    0.00
3      0      0      0       0.00    0.00    0.00    0.00
4      0      0      0       0       0.00    0.00    0.00
5      0      0      0       0       0       0.00    0.00
6      0      0      0       0       0       0       0.00
[1] "Decision Tree:"
       Year 0  Year 1  Year 2  Year 3  Year 4  Year 5
0      Wait    Wait    Wait    Invest  Invest  Invest
1      -       Wait    Wait    Wait    Invest  Invest
2      -       -       Wait    Wait    Wait    Wait
3      -       -       -       Wait    Wait    Wait
4      -       -       -       -       Wait    Wait
5      -       -       -       -       -       Wait
```

provides a graph (Fig. 6.12) that illustrates the likelihood of implementing (investing in) the project each year. Each bar shows the probability of "investing" in a particular year, compared to waiting. Using Monte Carlo simulation and 1000 iterations, we can say the chance of reaching a proper market condition to invest in this specific construction project would be the most in year 4, with the probability of nearly 37%. In other words, there is an approximately 37% chance that we would reach a node (a state in time) where the decision tree shows an "Invest" as an optimal decision. In each iteration, the model performs the investment valuation considering all the uncertainties and the risk-neutral probability of reaching a node based on the binomial mass function. This graph provides decision-makers with information that can be used to forecast the market and choose to wait or invest in a project at a particular time.

Many heavy infrastructure construction projects involve a public agency as an owner partnering with one or more private entities to undertake the project. This

6.4 Binomial Tree Model

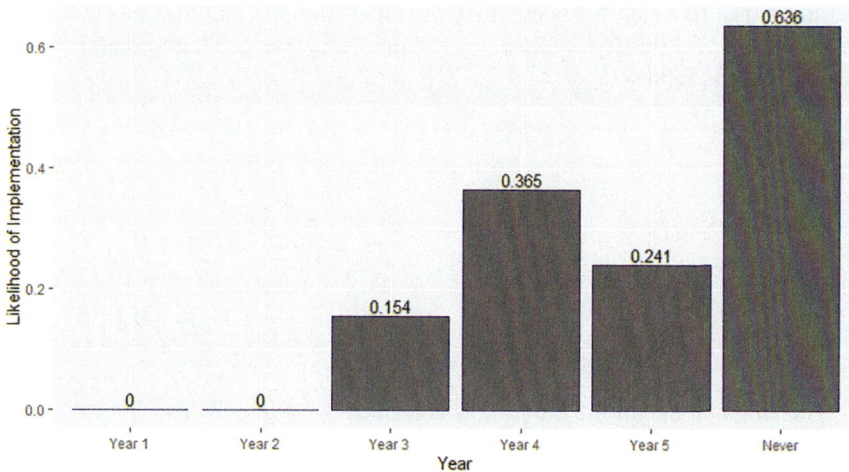

Fig. 6.12 Likelihood of investment as the optimal decision in each year

project delivery system is known as a public-private partnership (PPP). One common way of such agreements is where the private entity builds the infrastructure facility (e.g., tollway projects) and the public agency lends the facility for several years to the private party to collect revenues and compensate for costs. PPP has helped governments carry out more infrastructure projects despite their budgetary limitations. They use the expertise and financial resources of the private sector to maintain the old infrastructures and develop new ones. One of the most critical aspects of PPP agreements is risk sharing between the two parties. Real options analysis is a valuable instrument for measuring the amount of risk each party is willing to take in a PPP agreement.

Example 6.5

Suppose a local transportation agency intends to improve the regional mobility conditions of a city by building a four-lane tollway that connects two major parts of the city. A private firm has offered to finance and build the project and, in return, collect the toll revenue for 5 years. To share the risks, the transportation agency has agreed to set a minimum revenue guarantee (MRG) of $6.8 million a year. Based on the average annual daily traffic (AADT) data of a similar road, the expected number of cars taking the tollway is expected to be around 7,000,000 per year. Suppose the private firm is to collect $1 from each car taking the tollway and the variance of AADT is 10%. What would be the cost of the minimum revenue guarantee to the transportation agency? (Assume an interest rate of 5%.)

Solution The first step is to determine the input variables and then create a binomial tree to model traffic for future years. Considering the given information, input variables are as follows:

$X_0 = 7,000,000$ cars/year
$\Delta t = 1$ year
$r = 0.05$
$R_f = 1 + r = 1.05$
$K = e^{r\Delta t} = 1.0512$
$\sigma = 0.07$

$U = e^{\sigma\sqrt{\Delta t}} = 1.072$
$D = 1/U = 0.932$
$\pi_u = \dfrac{K - D}{U - D} = 0.8484$
$\pi_d = \dfrac{U - K}{U - D} = 1 - \pi_u = 0.1515$

The binomial tree of the traffic flow is created by setting the current state value of 7,000,000 and using $X(i,n) = X_0 U^{n-i} D^i$. In this example, we considered a 5-year contract to make the problem more explicable. However, the PPP concessions are usually more than 15 years in practice. One can utilize the method presented in this example and extend the problem to a 35-year contract. The binomial tree of the traffic flow for the next 5 years is shown in Table 6.4.

Considering the revenue of $1 a car, the annual revenue at each node can be calculated as $CF(i,n) = \$1 X(i,n)$; the result is the binomial tree of revenue cash flows presented in Table 6.5.

Table 6.4 Binomial tree of future traffic flow

X(i,n)	0	1	2	3	4	5
0	7,000,000.00	7,507,557.27	8,051,916.59	8,635,746.42	9,261,908.69	9,933,472.84
1		6,526,756.74	7,000,000.00	7,507,557.27	8,051,916.59	8,635,746.42
2			6,085,507.65	6,526,756.74	7,000,000.00	7,507,557.27
3				5,674,089.72	6,085,507.65	6,526,756.74
4					5,290,486.19	5,674,089.72
5						4,932,816.63
6						

Table 6.5 Binomial tree of estimated toll revenue cash flows with no MRG

CF(i,n)	0	1	2	3	4	5
0	–	7,507,557.27	8,051,916.59	8,635,746.42	9,261,908.69	9,933,472.84
1		6,526,756.74	7,000,000.00	7,507,557.27	8,051,916.59	8,635,746.42
2			6,085,507.65	6,526,756.74	7,000,000.00	7,507,557.27
3				5,674,089.72	6,085,507.65	6,526,756.74
4					5,290,486.19	5,674,089.72
5						4,932,816.63

6.4 Binomial Tree Model

Table 6.6 Binomial tree of estimated toll revenue cash flows with MRG application

CF'(i,n)	0	1	2	3	4	5
0	–	7,507,557.27	8,051,916.59	8,635,746.42	9,261,908.69	9,933,472.84
1		6,800,000.00	7,000,000.00	7,507,557.27	8,051,916.59	8,635,746.42
2			6,800,000.00	6,800,000.00	7,000,000.00	7,507,557.27
3				6,800,000.00	6,800,000.00	6,800,000.00
4					6,800,000.00	6,800,000.00
5						6,800,000.00

Table 6.7 Cost of the MRG based on different toll prices

Toll price	MRG	Toll price	MRG
$0.50	$11,876,784	$1.05	$17,189
$0.55	$10,120,418	$1.10	$6409
$0.60	$8,364,052	$1.15	$1690
$0.65	$6,607,687	$1.20	$251
$0.70	$4,904,171	$1.25	$121
$0.75	$3,381,311	$1.30	$24
$0.80	$2,084,147	$1.35	$9
$0.85	$1,099,184	$1.40	–
$0.90	$457,466	$1.45	–
$0.95	$214,597	$1.50	–
$1.00	$86,580		

Based on the MRG agreement, the transportation agency would recover the revenue deficit at each node with less than $6.8 million in cash flow. CF'(i,n) = Max { CF(i,n), $6,800,000} can replace the nodes with less than $6.8 million revenue cash flow. The binomial tree of the revenue cash flow with the MRG is presented in Table 6.6.

To calculate the cost of the MRG agreement, we can calculate the difference between the current market value of the cumulative revenue cash flows with and without introducing MRG. This can be done using the backward induction method explained earlier as:

$$V_{CF}(i,n) = CF(i,n) + \frac{\pi_u(i,n)V_{CF}(i,n+1) + \pi_d(i,n)V_{CF}(i+1,n+1)}{R_f} \quad (6.11)$$

where $V_{CF}(i,n)$ is the market value of the revenue cash flows at each node. Working back from the last period (year 5), the current market value of the revenue cash flows with and without MRG are $35,213,895 and $35,127,315, respectively. Therefore, the expected cost of the MRG option for the agency is $86,580.

Real options analysis can regulate the toll price based on the agency's budget. Table 6.7 and Fig. 6.13 show the cost of the MRG for the transportation agency based on different toll prices.

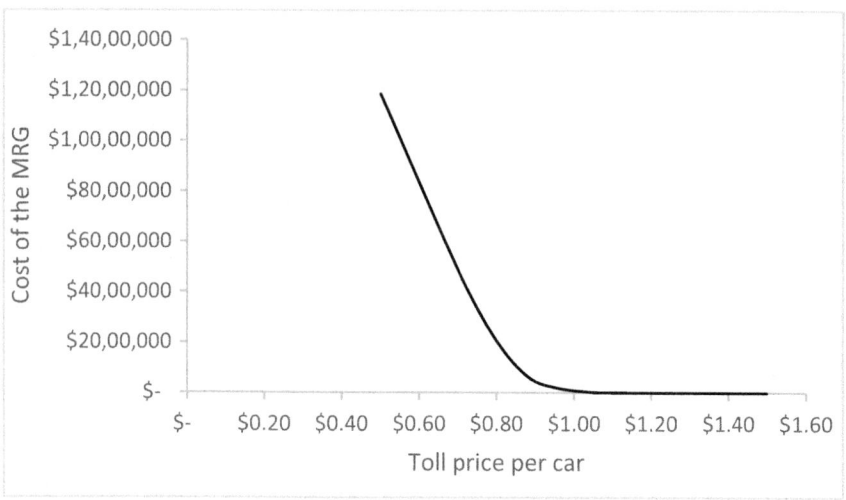

Fig. 6.13 The variation of the cost of MRG by different toll prices

6.5 Summary

Investment valuation of construction projects is subject to considerable uncertainties, such as variable costs and benefits. In this chapter, two principal mathematical and probabilistic methods (stochastic life-cycle cost analysis and real options analysis) were discussed to consider such uncertainties in the investment valuation of construction projects. The stochastic life-cycle cost analysis method captures the uncertainty of the investment valuation by incorporating the probability distribution of input variables. This method uses a Monte Carlo simulation technique to sample data from the input probability distributions and propagate them into the life-cycle cost analysis method. Then, the life-cycle cost of the project is presented as an empirical probability distribution or a cumulative probability distribution. Goodness-of-fit tests were used to characterize the resulting distributions parametrically. Further, a binomial tree model was discussed as a well-established real options analysis method to consider uncertainties in the investment valuation of construction projects. This method not only includes uncertainties regarding input variables but also accounts for flexibility in making decisions.

6.6 Exercise Problems

1. The mayor of a small city decides to rebuild an old recreational center in the center of the city. Due to limited funding sources, this investment requires a careful investment valuation considering construction cost uncertainties. Based on the data provided in Table 6.8 perform a stochastic LCCA of the project.

6.6 Exercise Problems

Table 6.8 Input variables of the two-lane road construction project

Input variable	Distribution type	Description
Engineering design	Normal	Mean = $500 K, SD = $200 K
Demolition of the old facility	Normal	$2 million, shape factor = 0.05, scale number = 0.2
Building construction	Triangular	a = $180 million, b = $250 million, c = $200 million
Landscaping	Triangular	a = $25 million, b = $60 million, c = $45 million
Annual maintenance and charges	Normal	Mean = $1 million, SD = $100 K

Table 6.9 Cost information on two alternative projects

Wind turbine power plant	Parameters	Distribution
Engineering design	μ = −$2 million SD = $500 K	Normal
Implementation cost	a = −$600 million b = −$350 million c = −$500 million	Triangular
Annual maintenance	a = −$2.5 million b = −$1 million c = −$2 million	Triangular
Annual revenue	μ = $100 million SD = $15 million	Normal
Coal power plant	*Cost and metrics*	*Description*
Engineering design	μ = −$3 million SD = $500 K	Normal
Implementation cost	a = −$600 million b = −$400 million c = −$450 million	Triangular
Annual maintenance	a = −$4 million b = −$3 million c = −$3.5 million	Triangular
Annual revenue	μ = $150 million SD = $10 million	Normal

Note: Assume this project has a service life of 50 years and an annual discount rate of 3%

2. The department of energy plans to build a new power plant to increase the capacity of the electricity system. There is a debate over choosing a wind turbine power plant or a conventional coal power plant. Based on the information in Table 6.9, perform a stochastic LCCA to identify the most profitable alternative.

   Consider a service life of 30 years and a 5% interest rate. Use the negative sign for costs and the positive sign for revenues.

3. A construction company has a 5-year window to invest in a shopping center project costing $200 million. It is assumed that the construction process would take 1 year and the current value of the completed complex is $400 million. Suppose the firm's interest rate is 8% and the volatility of the retail market is 5%. Considering these conditions, when would be the best time for the company to invest in the project?
4. Use the information in Example 6.4 and extend the project for 35 years. How much should the transportation agency budget for the MRG?
5. Considering the project in exercise problem 4, what is the maximum toll price that makes the project with no cost and risk-free for the agency? Hint: find the price that the MRG equals zero.

References

Arnold, T. B., & Emerson, J. W. (2011). Nonparametric goodness-of-fit tests for discrete null distributions. *The R Journal, 3*(2), 34–39.
Björk, T. (2009). *Arbitrage theory in continuous time*. Oxford University Press.
Burak Evrenosoglu, F. (2010). Modeling historical cost data for probabilistic range estimating. *Cost Engineering, 52*(5), 11.
Campbell, H. F., & Brown, R. P. (2003). *Benefit-cost analysis: Financial and economic appraisal using spreadsheets*. Cambridge University Press.
Copeland, T., & Antikarov, V. (2001). *Real options*. Texere.
Cox, J. C., Ross, S. A., & Rubinstein, M. (1979). Option pricing: A simplified approach. *Journal of Financial Economics, 7*(3), 229–263.
Darling, D. A. (1957). The Kolmogorov-Smirnov, Cramer-von Mises tests. *The Annals of Mathematical Statistics, 28*(4), 823–838.
Dixit, A. K., & Pindyck, R. S. (1994). *Investment under uncertainty*. Princeton University Press.
Gransberg, D. D., & Kelly, E. J. (2008). Quantifying uncertainty of construction material price volatility using Monte Carlo. *Cost Engineering, 50*(6), 14.
Guthrie, G. A. (2009). *Real options in theory and practice*. Financial Management Association.
Hull, J. C. (2017). *Options futures and other derivatives*. Pearson Education.
Jacquin, A. P., & Shamseldin, A. Y. (2007). Development of a possibilistic method for the evaluation of predictive uncertainty in rainfall-runoff modeling. *Water Resources Research, 43*(4), W04425.
Kashani, H., Ashuri, B., Shahandashti, S. M., & Lu, J. (2015). Investment valuation model for renewable energy systems in buildings. *Journal of Construction Engineering and Management, 141*(2), 04014074.
Pittenger, D., Gransberg, D. D., Zaman, M., & Riemer, C. (2012). Stochastic life-cycle cost analysis for pavement preservation treatments. *Transportation Research Record, 2292*(1), 45–51.
Raychaudhuri, S. (2008). Introduction to Monte Carlo simulation. In *2008 Winter simulation conference* (pp. 91–100). IEEE.
Rendleman, R. J., Jr. (1999). Option investing from a risk-return perspective. *Journal of Portfolio Management, 25*(5), 109–121.
Thomopoulos, N. T. (2012). *Essentials of Monte Carlo simulation: Statistical methods for building simulation models*. Springer Science & Business Media.
Walls, J., III, & Smith, M. R. (1998). *Life cycle cost analysis in pavement design-interim technical bulletin* (No. FHWA-SA-98-061). Federal Highway Administration.
Zahed, S. E., Shahandashti, S. M., & Diltz, J. D. (2020). Investment valuation of underground freight transportation systems under uncertainty. *Journal of Infrastructure Systems, 26*(3), 04020029.

Appendix A: Conventional Investment Valuation Techniques for Evaluating Construction Projects

This appendix reviews basic construction economics and conventional investment valuation techniques for evaluating construction projects. R codes are provided to implement these investment valuation techniques.

A.1 Cash Flow Analysis

The first step in conducting a cash flow analysis (CFA) is to identify all the cash flows that would occur throughout the life of the project or during the evaluation period. A construction project may involve many cash flows that can be categorized into two groups: "cash inflows" (earnings or benefits) and "cash outflows" (expenses or costs). Some of these cash flows frequently happen (recurring cash flows) during the life of a project, such as routine maintenance costs, monthly wages, or project revenues. Others only occur once or infrequently (nonrecurring cash flows or one-time cash flows) throughout the life of a project, such as design cost, construction cost, and the cost of purchasing material and equipment. In performing a cash flow analysis, it is essential to identify and include all the cash flows incurred in the life of a project from the very beginning to the end.

A.1.1 Project Cash Flows

The primary step in conducting a cash flow analysis is to define and characterize the project cash flows. A cash flow diagram (CFD) is used to graphically illustrate all the cash inflows and outflows of a project throughout its lifetime. In a cash flow stream, the project timeline is a horizontal line showing the life of the project in equal time intervals, and vertical arrows indicate the cash flows. While cash inflows are illustrated by arrows pointed upward, and above the timeline (positive), cash outflows or costs are shown by downward arrows placed below the horizontal timeline (negative)

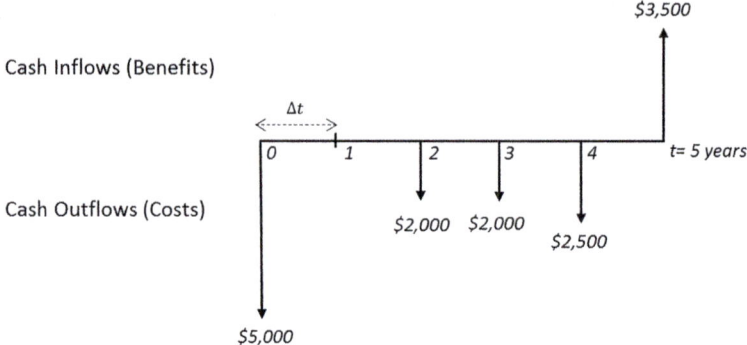

Fig. A.1 An example of a typical cash flow diagram

of the project. Figure A.1 illustrates an example of a cash flow diagram. It is always critical to determine the project lifetime (t) and the time intervals (Δt) before conducting a cash flow analysis and drawing a cash flow diagram. The project referred to in Fig. A.1 has a 5-year lifetime with 1-year time intervals. The vertical arrows represent the amount of the cash flows at the end of each time interval. This visualization technique helps us better understand the cash flows over the project life.

A cash flow diagram can also be presented as a bar chart. R Code A.1 can generate the cash flow diagram of the example shown in Fig. A.1 in the form of a bar chart (Fig. A.2).

R Code A.1
Cash flow diagram (CFD)

```
cf = c(-2000,-2000,-2500,3500)  #cf = Cashflow amounts
times = c(2,3,4,5)  #times = vector of the occurrence times for each cash flow
cf_t0 = -5000  #cf_t0 = Cashflow amounts
cfd(cf_t0 = cf_t0, cf=cf, times =times)  #cfd = function for cashflow diagram
```

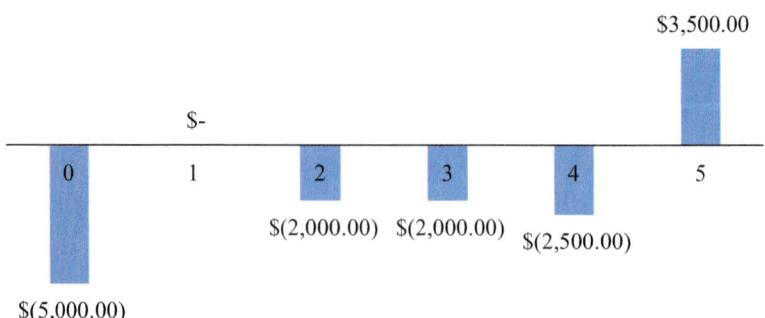

Fig. A.2 CFD in the form of a bar chart

Appendix A: Conventional Investment Valuation Techniques for Evaluating...

The vertical bars below and above the timeline represent cash outflows and inflows, respectively. The cash flow of a project at the end of each period (t) can be determined as the arithmetic summation of the cash flows incurred at that period. In other words, the amount of project cash flow at each period is the difference between the sum of all cash inflows and the sum of cash outflows at that period:

$$\text{Cash flow}_t = \sum(\text{Cash inflows})_t - \sum(\text{Cash outflows})_t \quad (A.1)$$

Suppose a construction project has the following CFD (Fig. A.3):

R Code A.2 can be used to calculate the cash flow of the project at each time interval.

Fig. A.3 CFD of a 5-year construction project

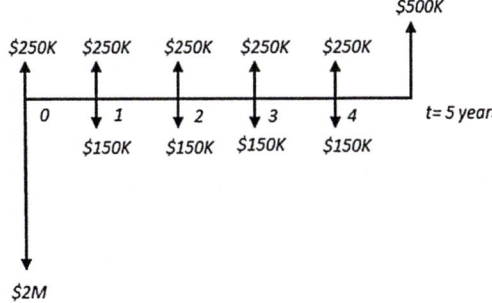

R Code A.2
Cash flow of the project at each time interval

```
cf = c(250,-150,250,-150,250,-150,250,150,500) #cf=Cashflow amounts
times = c(1,1,2,2,3,3,4,4,5) #times=vector of the occurrence times for each cash flow
cf_t0 = -2000 + 250 #cf_t0=cashflow at time 0 (Initial Cashflow) that can be positive or negative
cfd(cf_t0 = cf_t0, cf =cf, times = times) #cfd=function for cashflow diagram
```

Results

	Year	Cashflow
1	0	-1750
2	1	100
3	2	100
4	3	100
5	4	400
6	5	500

After building the cash flow diagram of an investment project, different engineering economics concepts, such as present value (PV), future value (FV), and equivalent uniform periodic values, can be calculated.

A.1.2 Net Present Value

Net present value (NPV) is a common investment valuation measure that shows the present value of a project considering all the costs and benefits (cash flows) during its lifetime. This measure can evaluate investment projects by comparing different investment alternatives. NPV of an investment project is obtained by discounting all the cash flows back to the present time and adding them all together. Equation (A.2) mathematically represents NPV calculation where r is the discount rate, t is the time period, and n is the life of the project.

$$NPV = \sum_{t=0}^{n} \frac{CF_t}{(1+r)^t}$$

(A.2)

R Code A.3 can be used to compute the NPV of cash flows. Assume a project has a cash flow diagram shown in Fig. A.4. Considering a 5% interest rate, the NPV of the project is calculated to be −$641.5 thousand. The negative NPV indicates that the project is not economically viable.

A.1.3 Future Value

Determining the future value of a cash flow stream could also be helpful in the investment valuation of construction projects. It helps estimate and compare the future values of various investments. The value of a cash flow stream at a particular time in the future is calculated by compounding the values of the cash flows increased due to the interest earned at the end of each period. For instance, a $100 cash flow would value $105 after a period with an interest rate of 5% per period. Going forward, the value of $100 cash flow after two periods is equal to $110.25, the same as the value of $105 after one period. In other words, the future value of a cash flow results from the interest earned on both principal (cash flow PV) and accrued interest. Equation A.3 is used to calculate the future value of cash flows:

$$FV = PV(1+r)^t$$

(A.3)

Appendix A: Conventional Investment Valuation Techniques for Evaluating...

R Code A.3
NPV of several cash flows

```
cf = c(100,300)    #cf= Cashflow amounts
times = c(3,2)     #times= vector of the occurrence times for each cash flow
cf_t0 = -1000      #cf_t0= cashflow at time 0 (Initial Cashflow) that can be positive
or negative
npv(cf_t0 = cf_t0, cf = cf, times = times, r = 0.05)  #i= interest rate per period,
NPV=the function to calculate the net present value
```

Results

```
[1] -641.5074
```

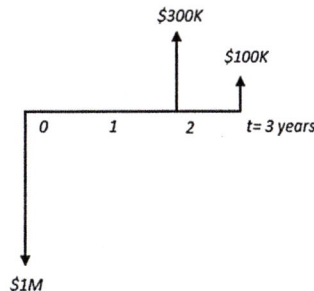

Fig. A.4 CFD of a 3-year project

R Code A.4 can be used to convert present and future values. For instance, the future value of a $50 million investment would be around $60.8 million after 5 years, considering a 4% rate of return. On the other hand, if the rate of return on an investment is 4%, we should invest around $41.1 million today to have a $50 million value after 5 years.

R Code A.4
Calculating the future value of several cash flows

```
tvm(pv=50, n=5, r=.04) #TVM= the function of time value of money for calculating
future and present values, pv= present value
tvm(fv=50,n=5,r=.04) #fv= future value

# input "pv" to calculate the future value or input "fv" to calculate the present
value.
```

Results

| FV | 60.83265 |
| PV | 41.09636 |

A.1.4 Equivalent Uniform Values

It is sometimes helpful to convert cash flows into equivalent uniform values distributed throughout the life of the project. This technique is most suitable for comparing projects with unequal lifetimes or evaluating the required periodic (annual or monthly) budget for developing and owning (operating and maintaining) that project. Equations (A.4) and (A.5) can be used to transform all the cash flows into equal uniform values and add them together. Since many investment valuations are performed using yearly time periods ($\Delta t = 1$ year), equivalent uniform values are substituted by other terms such as "annuity" and "equivalent uniform annual costs" (EUAC)."

$$A = PV \left[\frac{r(1+r)^n}{(1+r)^n - 1} \right] \tag{A.4}$$

$$A = FV \left[\frac{r}{(1+r)^n - 1} \right] \tag{A.5}$$

R Code A.5 determined the equivalent uniform annual cost of a cash flow. This example shows that if we borrow $1000 today with a 5% interest rate, we should pay approximately $123.33 annually for 10 years to pay it back. Likewise, the value of investing $79.5 a year with an interest rate of 5% would be $1000 after 10 years. This example clearly shows the difference between lending and borrowing money.

Appendix A: Conventional Investment Valuation Techniques for Evaluating...

R Code A.5
Calculating the EUAC of a cash flow

```
# euv = the function of equivalent uniform values, n = number of payments/periods,
a = amount of the first payment, q = payment increment amount per period, i =
nominal interest frequency convertible ic times per year, ic= interest conversion
frequency per year, pf = the payment frequency- number of payments per year, imm
= option for annuity immediate or annuity due, default is immediate (TRUE)

euv(pv=1000, fv=NA, n=10, a=NA, q=0, r=.05, ic=1, pf=1, imm=FALSE)   # annuity,
having present value
euv(pv=NA, fv=1000, n=10, a=NA,q=0, r=.05, ic=1, pf=1, imm=TRUE)   # annuity,
having future value
```

Results

```
Level Annuity
PV 1000.0000
FV 1628.8946
A 123.3377
Q 0.0000
Eff Rate 0.0500
Years 10.0000

Level Annuity
PV 613.91325
FV 1000.00000
A 79.50457
Q 0.00000
Eff Rate 0.05000
Years 10.00000
```

Example A.1

A construction company spends $250 thousand per year renting a crane for its projects. Suppose the owner decides to evaluate purchasing a crane for the company. The initial cost of the crane is $2 million. The cost of maintaining this crane to stay operational is $150 thousand per year. Is this investment beneficial if the crane could be salvaged at $500 thousand after its 5-year service life? Assume the discount rate to be 4%.

Solution NPV analysis can be used to compare the two alternatives and determine whether owning a crane is a sound investment. R Code A.6 can be used to conduct NPV analysis for the two alternatives (purchasing vs. renting the equipment).

R Code A.6
NPV analysis of two alternative scenarios

```
# Define the cash flow of each project as the arrow c.
# number of cashflows for all the projects should be the same. Use zero for years
without any cash flows.
project1 = c(-2000, -150, -150, -150, -150, 350)    # Cashflows for project1
project2 = c(-250, -250, -250, -250, -250, -250)    # Cashflows for project2
npv_ia(project1 = project1, project2 = project2, project3 = NA, r = 0.04, cf_t0
= TRUE) # npv_investment_assessment = the function of NPV analysis that could
compare up to three projects together.
# cf_t0 = cash flow at time zero. Set to (TRUE) if you have any cash flows at
time zero. Otherwise set to (FALSE).
# r = Discount rate
```

Results

Metric	project1	project2	project3
1 Net Present Value (NPV), USD mln	-2256.81	-1362.956	NA

The results show that renting a crane provides a greater NPV and is a more beneficial investment than buying a crane for this company.

This investment decision case can also be evaluated by transforming all the costs into equivalent uniform annual costs and comparing the annual costs of two scenarios. Run the R codes provided in R Code A.7 to see the results. The alternative with the lowest EUAC would be the most economical: the renting option.

R Code A.7
EUAC analysis of two alternative scenarios

```
#Define the cash flow of each project as the arrow c.
# number of cashflows for all the projects should be the same. Use zero for years
without any cash flows.
project1 = c(-2000, -150, -150, -150, -150, 350)    # Cashflows for project1
project2 = c(-250, -250, -250, -250, -250, -250)    # Cashflows for project2
euv_ia(project1 = project1, project2 = project2, project3 = NA, r = 0.04, cf_t0
= TRUE) # euv_investment_assessment = the function of EUAC analysis that could
compare up to three projects together.
# cf_t0= cash flow at time zero. Set to (TRUE) if you have any cash flows at time
zero. Otherwise set to (FALSE).
# r= Discount rate
```

Results

```
Metric          project1    project2    project3
1 EUV           -506.9407   -306.1568   NA
```

The United States Office of Management and Budget (USOMB) publishes recommended discount rates annually, which can be used for public projects.

A.2 Life-Cycle Cost Analysis

Life-cycle cost analysis (LCCA) is one of the most effective investment valuation techniques enabling decision-makers to rigorously evaluate and compare the life-cycle cost of different projects with the same objective and identify the most cost-effective alternative (FHWA, 2002). Construction companies have frequently used LCCA to identify the most cost-effective alternatives for different aspects of projects, such as design, construction method, and material selection. For example, LCCA can be used to compare two alternative construction methods of opencut or mechanized tunneling for constructing a 100-mile underground freight transportation (UFT) tunnel. Likewise, the life-cycle cost (LCC) of different heating, ventilation, and air-conditioning (HVAC) systems can be compared to identify the most cost-effective option to be used in a large multipurpose building.

Cash flow analysis is the basis of LCCA. To conduct an LCCA, all the costs involved in the life cycle of each alternative option should be identified, and a cash flow analysis should be conducted for each alternative. Different measures, such as NPV and EUAC, can be used to compare alternatives and identify the most cost-effective method.

Example A.2

A local transportation agency decides to build a new highway to expand the transportation capacity between two cities. A transportation consultant company has proposed two design alternatives for the highway pavement: a rigid concrete pavement and a flexible asphalt pavement. The consulting firm has also provided the cost estimation of the alternatives, as shown in Table A.1. Which of the alternatives is the most cost-effective option? (Assume a discount rate of 5%.)

Table A.1 Life-cycle cost estimation of the two pavement alternatives

	Concrete pavement	Asphalt pavement
Initial construction cost	$400 million	$250 million
Annual maintenance cost	$20 million	$30 million
Expected service life	40 years	25 years

Solution Since the service life of the two alternatives is not the same, the EUAC measure is recommended to compare them and find the option with the lowest annual cost. Run R Code A.8 to see the results of this analysis.

R Code A.8
LCCA analysis of two alternative scenarios

```
NPV_Concrete_Pavement = npv(cf_t0=400000000, cf = rep(20000000, length.out = 40),
times=1:40, r=.05)
NPV_Asphalt_Pavement = npv(cf_t0=250000000, cf = rep(30000000, length.out = 25),
times=1:25, r=.05)
EUV_Concrete_Pavement = euv(pv=NPV_Concrete_Pavement, fv=NA, n=40, a=NA, q=0,
r=.05, ic=1, pf=1, imm=TRUE)
EUV_Asphalt_Pavement = euv(pv= NPV_Asphalt_Pavement, fv=NA, n=25, a=NA, q=0,
r=.05, ic=1, pf=1, imm=TRUE)
sprintf("NPV for Concrete Pavement and Asphalt Pavement is %1.0f and %1.0f,
respectively", NPV_Concrete_Pavement,NPV_Asphalt_Pavement)
sprintf("EUV for Concrete Pavement and Asphalt Pavement is %1.0f and %1.0f,
respectively", EUV_Concrete_Pavement[3], EUV_Asphalt_Pavement[3])
```

Results

[1] "NPV for Concrete Pavement and Asphalt Pavement is 743181727 and 672818337, respectively"

[2] "EUV for Concrete Pavement and Asphalt Pavement is 43311264 and 47738114, respectively"

Results show that although the initial cost of concrete pavement is higher than that of asphalt pavement, the equivalent annual cost will be lower than the annual cost of asphalt pavement in the life cycle.

A.3 Life-Cycle Benefit-Cost Analysis

Life-cycle benefit-cost analysis (LCBCA) is a systematic decision-making approach that helps us evaluate projects by considering both their costs and benefits. LCBCA usually includes not only direct benefits that have monetary values, such as revenues, but also indirect benefits, such as cost avoidances, environmental benefits, and public satisfaction. This analysis can be used to compare various projects or different alternatives. Unlike LCCA, LCBCA can be used for comparing projects with different outcomes and benefits. LCBCA helps decision-makers select the best alternative with the highest benefits compared to its costs. Furthermore, LCBCA has historically been regarded as the most impartial and credible outcome of feasibility studies in the construction industry, notably for infrastructure projects.

Suppose local transportation officials want to increase the freight transportation capacity between two cities of A and B. They have different alternatives, such as expanding the existing highway with additional lanes, building a new highway designated for only freight vehicles, building freight rail lines, and building an underground freight transportation system. Although all the alternatives satisfy the ultimate goal of increasing the freight transportation capacity between the two cities, they have many other different economic impacts (both in terms of cost and benefit). For example, building an underground freight transportation system would have many social benefits, such as reducing environmental pollution and the number of accidents that should be accounted for in the evaluation.

The general notion of benefit-cost analysis is to identify the best economic alternative by comparing benefits to costs. Common economic measures, including the benefit-cost ratio (BCR), net present value (NPV), internal rate of return (IRR), and breakeven point, can be used to conduct a benefit-cost analysis.

A.3.1 Benefit-Cost Ratio (BCR)

The benefit-cost ratio is the relative value of the project benefits to its costs. This value is usually calculated as the sum of the discounted value of the benefits (present value) divided by the sum of the discounted value of costs. Equation (A.6) shows the mathematical formulation of the benefit-cost ratio, where B_t and C_t are the sum of the benefit and cost, respectively, at time t.

$$BCR = \frac{\sum_{t=0}^{n} \frac{B_t}{(1+r)^t}}{\sum_{t=0}^{n} \frac{C_t}{(1+r)^t}} \quad (A.6)$$

The proposed project is economically desirable if the benefit-cost ratio is greater than or equal to 1.0. Furthermore, the project (or alternative) with the highest benefit-cost ratio is the most economical choice.

R Code A.9 can be used to determine the benefit-cost ratio of a project with various benefits and costs. Assume a project has a cash flow diagram depicted in Fig. A.5.

The benefit-cost ratio of the project is 1.8, which shows the project would be profitable.

R Code A.9
Calculating the BCR of a project

```
# Use positive values for cash inflows and negative values for cash outflows!!!
cif_t0 = 1000   # cif_t0 = initial cash inflow
cif = c(2000,3000,1000) # cif = vector of cash inflows
cif_times = c(1,2,3) # cif_times = vector of cash inflow occurrence times
cof_t0 = -1000  # cof_t0 = initial cash outflow
cof = c(-100,-3000) # cof = vector of cash outflows
cof_times = c(1,3) # cof_times = vector of cash outflow occurrence times
r= 0.1 # r= interest/discount rate
bc_ratio(cif_t0 = cif_t0, cif = cif, cif_times = cif_times, cof_t0 = cof_t0, cof = cof, cof_times = cof_times, r= r) # benefit_cost_ratio = function for calculating BCR
```

Results

```
[1] 1.808401
```

Appendix A: Conventional Investment Valuation Techniques for Evaluating...

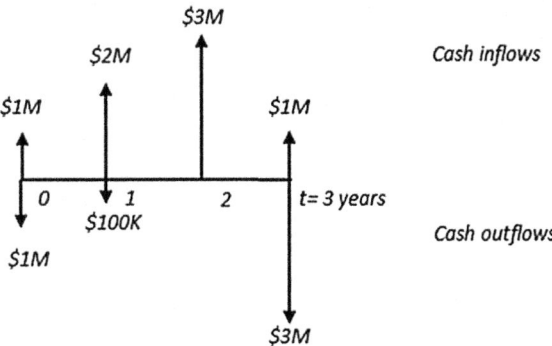

Fig. A.5 CFD of a 3-year construction project

A.3.2 Net Present Value (NPV)

The net present value of each alternative is calculated using Eq. (A.7). A positive NPV shows the benefits of the project outweigh the costs and indicates the project is economically sound. Likewise, a project with the highest NPV is the best economic choice between alternatives.

$$NPV = \sum_{t=0}^{n} \frac{B_t - C_t}{(1+r)^t} \quad (A.7)$$

R Code A.3 can also be used to calculate the NPV of a project considering both costs and benefits.

A.3.3 Internal Rate of Return (IRR)

The internal rate of return can be used to show how much a project is profitable. The internal rate of return is a discount rate that brings the NPV of the project to zero. If the rate of return exceeds the adopted market discount rate, the project is economically viable. In the case of alternative projects, a project with the highest internal rate of return is the most beneficial choice. Equation (A.8) is a simple mathematical representation of IRR.

$$NPV = 0 = \sum_{t=0}^{n} \frac{B_t - C_t}{(1+IRR)^t} \quad (A.8)$$

R Code A.10 can be used to calculate the internal rate of return of a project. Assume investment in building a facility with an initial construction cost of $200 million that is expected to generate $150 million in revenue for 4 years and be sold for $350 million after 5 years. The internal rate of return of the project is calculated to be 75%, which indicates a significant profit margin.

R Code A.10
Calculating the IRR of a project

```
project1 = c(-200, 150, 150, 150, 150, 350)   # Cashflows for project1
irr_ia(project1           =             project1,        cf_t0         =          TRUE)
# irr_investment_assessment is the function of calculating the IRR of a project.
# cf_t0 = cash flow at time zero. Set to (TRUE) if you have any cash flows at
time zero. Otherwise set to (FALSE).
```

Results

```
        project1
IRR     0.7500016
```

A.3.4 Breakeven Analysis

Breakeven analysis is a widespread decision-making technique for new investments and ongoing projects. This analysis compares the cumulative amount of cash inflows and outflows of a project along its lifetime to identify the point that the project returns all the investment. In other words, breakeven analysis reveals the time that the project turns into the profitability phase. This point in time of the project is called the "breakeven point" or recovery point. The period that it takes to reach the breakeven point is known as the "payback period."

The payback period can be used as a benchmark to choose between alternative projects and scenarios. A project with a shorter payback period is always more attractive to investors. Breakeven analysis plays a significant role in the financial planning of projects. Breakeven analysis helps managers carry out financial planning and decide how to distribute funds throughout the life of the project.

To conduct a breakeven analysis, we should first identify all the cash flows of the project and perform a cash flow analysis. The breakeven point is the point when the sum of the discounted costs (cash outflows) becomes equal to the sum of the discounted benefits (cash inflows), as shown in Eq. (A.9). The breakeven point can be found by solving this equation for n.

$$\sum_{t=0}^{n} \frac{(\text{Cash inflows})_t}{(1+r)^t} = \sum_{t=0}^{n} \frac{(\text{Cash outflows})_t}{(1+r)^t} \qquad (A.9)$$

Example A.3

A state department of transportation plans to develop an underground freight transportation system as a novel approach to increase the capacity of freight transportation between two major cities. Although the capital cost of implementing such a system is substantial, it has many social and indirect benefits that would significantly improve the quality of life in the region. Suppose the state department of transportation intends to enter a public-private partnership contract with a private company that requires the company to fund and build the system and, in return, the company can operate and collect revenue for 30 years. Table A.2 provides the monetary value of all costs and benefits of the system. Use breakeven analysis to find the payback period of the proposed underground freight transportation system (assume a discount rate of 5%).

Table A.2 Life-cycle cost valuation of the proposed system

UFT system benefits	Value
Environmental pollution reduction	$200,000,00/year
Traffic congestion reduction	$45,000,000/year
Infrastructure damage cost reduction	$11,000,000/year
Other social benefits	$70,000,000/year
Shipment revenue	$798,437,500/year
UFT system costs	Value
Construction cost	$5,000,000,000
Equipment costs	$2,000,000,000
Equipment maintenance	$20,000,000/year
Facility maintenance	$10,000,000/year
Operation cost	$100,000,000/year

Solution Using the data provided in Table A.2 and R Code A.11, one can conclude that the project yields a return on the investment near the end of the eighth year (8.89) and enter a profitability zone.

R Code A.11
Calculating the payback period of the proposed UFT project

```r
cif_t0 = 0 # No inflow at t0
cif = rep((798437500 + 200000000 + 45000000 + 11000000 + 70000000), length.out = 30) # Recurring cash inflows for 30 years
cof_t0 = 5000000000 + 2000000000 #initial investment at t0
cof = rep((20000000 + 10000000 + 100000000), length.out = 30) # Recurring cash outflows for 30 years
cashflow_t0 = cif_t0 - cof_t0
cashflows = cif - cof
project1 = append(cashflow_t0 , cashflows)
dpbp_ia(project1 = project1,r = 0.05,cf_t0 = TRUE)
# dpbp_investment_assessment is the function of calculating the duration of the payback period of a project. cf_t0 = cash flow at time zero. Set to (TRUE) if you have any cash flows at time zero. Otherwise set to (FALSE).
```

Results

	Metric	project1	project2	project3
1	Discounted Payback Period, years	8.893476	NA	NA

Figure A.6 is the graphical representation of the payback period. This graph shows that the return on investment for this project starts in about 9 years.

Breakeven analysis has been widely utilized in various industries to make many strategic decisions, such as operational level, pricing, and logistics. A construction company can use breakeven analysis to negotiate with the owner on the costs to secure a certain amount of profit and recover all the expenses. Also, a railroad company can use breakeven analysis to ascertain the minimum shipment price or the minimum level of operation that recovers all the costs of expanding its railroad network in a certain period of time.

Example A.4
Consider the information provided in Example A.3. Use BCR, NPV, and IRR measures and perform an LCBCA to determine if the proposed project is economically feasible.

Appendix A: Conventional Investment Valuation Techniques for Evaluating... 141

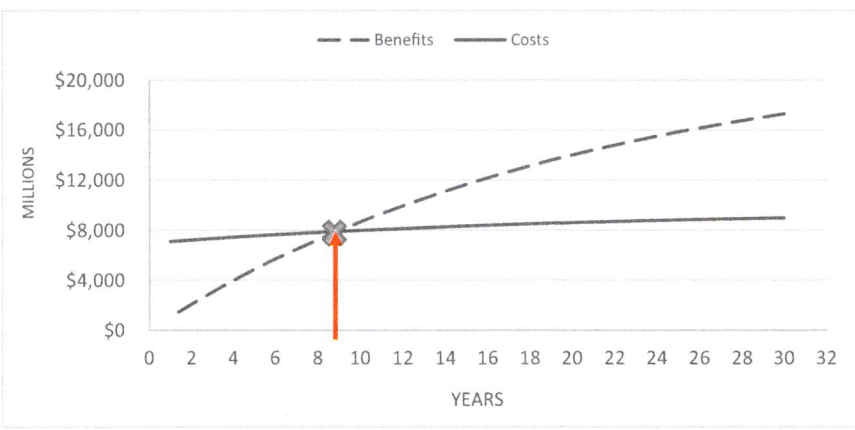

Fig. A.6 Breakeven point of the studied UFT project

Solution First, a cash flow analysis helps understand the amount and distribution of the cash flows throughout the life of the project. Then, the required economic measures (BCR, NPV, and IRR) of the project are calculated using the codes presented above. R Code A.12 can be used to solve the problem.

R Code A.12
LCBCA of the proposed UFT project

```
# this code calculates the NPV, IRR, UAV and the duration of the payback period
(DPBP) of the project.
cif_t0 = 0 #No inflow at t0
cif = rep((798437500 + 200000000 + 45000000 + 11000000 + 70000000), length.out =
30) #Recurring cash inflows for 30 years
cof_t0 = 5000000000 + 2000000000 #initial investment at t0
cof = rep((20000000 + 10000000 + 100000000), length.out = 30) #Recurring cash
outflows for 30 years
cashflow_t0 = cif_t0 - cof_t0
cashflows = cif - cof
project1 = append(cashflow_t0 , cashflows)
cif_times = rep(1, length.out = 30)
cof_times = rep(1, length.out = 30)
ia(project1 = project1, r = 0.05, cf_t0 = TRUE)
BCR = bc_ratio(cif_t0 = cif_t0, cif = cif , cif_times = cif_times, cof_t0 =
cof_t0, cof = cof, cof_times = cof_times, r= 0.05)
sprintf("The Benefit Cost Ratio of the Project is: %.3f", BCR)
```

Results

```
Metric  project1        project2    project3
NPV     8.286942e+09    NA          NA
IRR     1.392171e-01    NA          NA
DPBP    8.893476e+00    NA          NA
EUV     5.390775e+08    NA          NA
"The Benefit Cost Ratio of the Project is: 2.999"
```

A.4 Sensitivity Analysis and Cost Contingencies

Successful investment analysis requires a rigorous evaluation of input variables and their impacts on the results. Sensitivity analysis is a conventional technique to assess the vulnerability of the results to specific changes in the input variables. Since cost estimation of construction projects is usually subjected to high-cost contingencies, a sensitivity analysis is necessary to reach a robust decision. Sensitivity analysis is performed by changing an input variable to a certain degree while keeping all the other inputs constant. The percent change in the final results is the sensitivity of the results to that specific input variable. This analysis should be conducted for different input variables to identify key variables with the most level of influence on the investment valuation results. The cost contingencies or the level of change in the input variables is usually determined based on experience, historical data, or predetermined guidelines. Identifying key input variables helps decision-makers to be more meticulous in estimating those variables. It also helps them to be prepared for a range of possible outcomes.

Let's consider the project presented in Example A.3 and evaluate the sensitivity of the benefit-cost ratio to the following input variables: construction cost, maintenance cost, equipment cost, shipment revenue, reduction in environmental impact, and reduction in traffic congestion. Let's also assume a ±50% level of contingency for the input variables. Using sensitivity analysis, we can identify the most critical benefit and cost components of the proposed UFT system. Figure A.7 illustrates the results of this analysis indicating the construction cost as the most vital variable affecting the benefit-cost ratio of the project (the line with the highest slope). Thus, it is essential for the design and construction company to put every effort into keeping the construction cost within the budget.

A.5 Summary

This chapter presents the most common deterministic investment valuation approaches to evaluate construction-related projects. We showed how to first set up your investment valuation case using cash flow diagrams and how to perform a life-cycle cost analysis of the project. This analysis provides information, such as NPV, future value, and equivalent uniform value of project cash flows, considering its life

Appendix A: Conventional Investment Valuation Techniques for Evaluating… 143

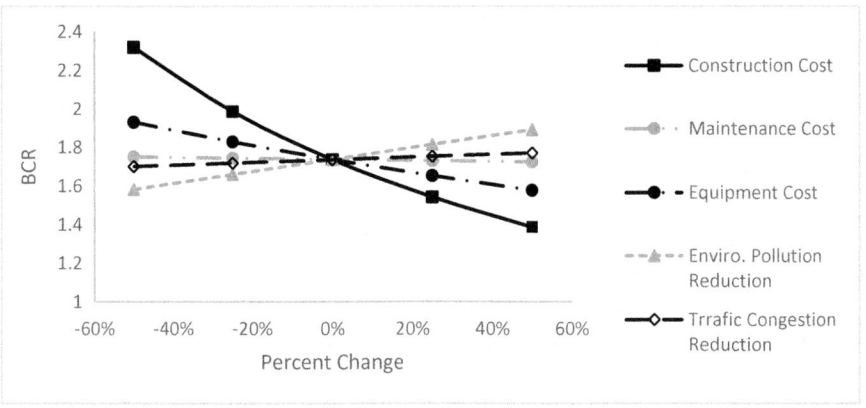

Fig. A.7 Results of sensitivity analysis of the studied UFT system

cycle. This information can be used for making investment decisions, such as comparing different scenarios or alternative projects and preparing financial plans to execute a project. Moreover, we presented a life-cycle benefit-cost analysis method that considers both the benefit and cost of construction projects at the same time. This method provides more insights into the projects, such as the benefit-cost ratio, the internal rate of return, and the payback period of the project. This information helps managers make better and more informed decisions about investing in construction projects and comparing alternatives. Finally, we presented the sensitivity analysis approach to assess the robustness of the deterministic investment analysis results. This method can identify key input parameters and determine how much they influence our investment analysis results.

A.6 Exercise Problems

1. A local transportation agency decides to develop a toll road to improve the accessibility of a major city. Developing the toll road would cost $1 billion and 4 years to complete. Suppose the construction cost is distributed evenly over the construction years. The concessionaire (private investor) would operate the facility for 20 years based on the contract. The revenue from toll collection is estimated to be $100 million per year and starts from year 5, 1 year after the construction project is finished. Moreover, the operation and maintenance of the facility would cost the private party $20 million per year. Draw the CFD of the project cash flows.
2. Considering the project presented in problem 1, what is the total cash flow of the project each year?

3. What is the net present value of the investment project described in problem 1, considering a 5% interest rate? Is investment in this project economically viable?
4. Considering the project presented in problem 1, for budget planning, what would be the average cash flow of the project considering the interest rate of 5%? (Hint: calculate the EUAC of the project.)
5. The value of a commercial parcel of land in the downtown of a major city is increasing by a rate of 10% per year. If the land is worth $50 million today, (a) what would it worth after 10 years? Suppose a venture capitalist offers the owner $60 million to buy the land. (b) If the market interest rate is 5%, is it better for the owner to sell the land or wait for 10 years and sell it after 10 years?

References

FHWA. (2002). *Life-cycle cost analysis primer, Office of Asset Management*. Federal Highway Administration.

Appendix B: R and R Package Installation and Helpful Communities

- Website for R installation: http://cran.r-project.org/
- R package archive: https://cran.r-project.org/web/packages/index.html
- R package documentation: https://www.rdocumentation.org/; https://rdrr.io/
- RPubs by RStudio: https://rpubs.com/
- RStudio community: https://community.rstudio.com/
- RStudio blog: https://blog.rstudio.com/
- R-bloggers: https://www.r-bloggers.com/
- GitHub for R users: https://github.com/topics/r

Appendix C

Not seasonally adjusted federal Highway Construction Spending (HCS) from 2003 to 2020

	Jan	Feb	Mar	Apr	May	Jun	Jul	Aug	Sep	Oct	Nov	Dec
2003	28	31	29	42	58	80	72	77	143	67	56	41
2004	55	33	51	84	84	99	108	113	171	69	47	29
2005	28	21	33	34	47	61	76	98	109	37	56	33
2006	34	35	43	35	49	54	41	54	55	43	59	34
2007	32	26	30	54	57	67	85	115	122	84	63	57
2008	36	37	43	56	62	62	78	87	87	78	58	54
2009	45	42	40	45	54	73	99	86	108	68	96	46
2010	34	48	52	68	79	105	150	156	186	96	74	61
2011	43	36	46	47	74	76	87	112	163	105	74	56
2012	38	37	49	42	64	60	67	74	61	58	74	47
2013	34	31	40	51	63	65	83	81	89	49	30	26
2014	10	14	20	23	37	41	70	63	78	53	33	31
2015	16	22	17	32	36	56	76	75	68	53	47	46
2016	35	51	56	42	57	56	68	105	137	75	70	90
2017	72	62	89	75	85	92	108	105	162	107	80	74
2018	63	60	67	68	81	82	70	102	152	101	94	75
2019	82	79	105	130	164	144	153	138	179	108	92	75
2020	57	56	72	78	92	121	116	117	134	100	114	112

Appendix D

National Highway Construction Cost Index (NHCCI) from 2003 to 2018

	Qtr1	Qtr2	Qtr3	Qtr4
2003	1.000	1.010	1.024	1.022
2004	1.046	1.101	1.143	1.149
2005	1.241	1.281	1.372	1.412
2006	1.449	1.521	1.618	1.553
2007	1.564	1.561	1.538	1.514
2008	1.569	1.644	1.785	1.627
2009	1.500	1.440	1.429	1.403
2010	1.442	1.438	1.447	1.430
2011	1.457	1.501	1.541	1.541
2012	1.577	1.627	1.596	1.607
2013	1.591	1.624	1.645	1.593
2014	1.628	1.670	1.735	1.694
2015	1.720	1.705	1.706	1.663
2016	1.631	1.678	1.680	1.653
2017	1.617	1.685	1.734	1.662
2018	1.675	1.752	1.845	1.873

Appendix E

Price Trend Index for Iowa Highway Construction (IHC) from 2007 to 2018

	Qtr1	Qtr2	Qtr3	Qtr4
2007	210.5	253.1	221.3	264.2
2008	225.6	256.9	291.1	281.3
2009	248.7	247.4	253.8	273.5
2010	234.2	247.0	233.2	227.2
2011	248.7	249.6	238.8	305.5
2012	265.5	306.4	289.6	273.8
2013	258.5	279.9	291.0	283.0
2014	291.3	288.9	308.3	326.4
2015	361.3	331.3	308.1	270.9
2016	267.0	290.1	274.2	294.6
2017	286.6	286.4	300.9	274.3
2018	287.6	289.8	343.1	324.7

Appendix F

California Construction Cost Index (CCCI) from 2006 to 2019

	Jan	Feb	Mar	Apr	May	Jun	Jul	Aug	Sep	Oct	Nov	Dec
2006	4620	4604	4597	4600	4599	4593	4609	4616	4619	4867	4891	4877
2007	4869	4868	4871	4872	4886	4842	4849	4851	4942	4943	4978	4981
2008	4983	4983	4999	5004	5023	5065	5135	5142	5194	5393	5375	5322
2009	5309	5295	5298	5296	5288	5276	5263	5265	5264	5259	5259	5262
2010	5260	5262	5268	5270	5378	5394	5401	5401	5381	5591	5599	5596
2011	5592	5624	5627	5636	5637	5643	5654	5667	5668	5675	5680	5680
2012	5683	5683	5738	5740	5755	5754	5750	5778	5777	5780	5779	5768
2013	5774	5782	5777	5786	5796	5802	5804	5801	5802	5911	5903	5901
2014	5898	5896	5953	5956	5957	5961	5959	5959	5959	5969	5981	5977
2015	6073	6077	6069	6062	6069	6055	6055	6055	6113	6114	6109	6108
2016	6106	6132	6248	6249	6240	6238	6245	6244	6267	6343	6344	6373
2017	6373	6373	6373	6461	6455	6470	6474	6620	6620	6596	6596	6596
2018	6596	6596	6596	6596	6596	6598	6643	6613	6674	6679	6679	6684
2019	6684	6700	6616	6841	6852	6854	6854	6823	6814	6851	6895	–

Appendix G

Monthly Highway Construction Spending (HCS) from 2003 to 2019

	Jan	Feb	Mar	Apr	May	Jun	Jul	Aug	Sep	Oct	Nov	Dec
2003	113	113	122	90	67	70	73	71	85	90	143	98
2004	69	81	81	67	58	59	62	74	80	107	162	105
2005	108	92	85	75	89	62	72	90	70	75	137	105
2006	68	56	57	42	56	51	35	46	47	53	68	75
2007	76	56	36	32	17	22	16	31	33	53	78	63
2008	70	41	37	23	20	14	10	26	30	49	89	81
2009	83	65	63	51	40	31	34	47	74	58	61	74
2010	67	60	64	42	49	37	38	56	74	105	163	112
2011	87	76	74	47	46	36	43	61	74	96	186	156
2012	150	105	79	68	52	48	34	46	96	68	108	86
2013	99	73	54	45	40	42	45	54	58	78	87	87
2014	78	62	62	56	43	37	36	57	63	84	122	115
2015	85	67	57	54	30	26	32	34	59	43	55	54
2016	41	54	49	35	43	35	34	33	56	37	109	98
2017	76	61	47	34	33	21	28	29	47	69	171	113
2018	108	99	84	84	51	33	55	41	56	67	143	77

© The Editor(s) (if applicable) and The Author(s), under exclusive license to Springer Nature Switzerland AG 2023
M. Shahandashti et al., *Construction Analytics*,
https://doi.org/10.1007/978-3-031-27292-9

Appendix H

Monthly consumer price index (CPI) from 2003 to 2018

	Jan	Feb	Mar	Apr	May	Jun	Jul	Aug	Sep	Oct	Nov	Dec
2003	266.9	269	270.6	269.9	269.5	269.9	270.2	271.2	272	271.8	271	270.7
2004	272.1	273.6	275.3	276.2	277.7	278.6	278.3	278.4	278.9	280.4	280.6	279.6
2005	280.1	281.7	283.9	285.8	285.6	285.7	287	288.5	292	292.6	290.3	289
2006	291.3	291.9	293.6	296.0	297.4	298.1	298.9	299.6	298.1	296.5	296.0	296.5
2007	297.4	299.0	301.7	303.6	305.5	306.1	306.0	305.5	306.3	306.9	308.8	308.6
2008	310.1	311.0	313.7	315.6	318.2	321.5	323.2	321.9	321.4	318.2	312.1	308.8
2009	310.2	311.7	312.5	313.3	314.2	316.9	316.4	317.1	317.3	317.6	317.8	317.3
2010	318.3	318.4	319.7	320.3	320.5	320.2	320.3	320.7	320.9	321.3	321.5	322.0
2011	323.6	325.2	328.3	330.5	332.0	331.7	332.0	332.9	333.4	332.7	332.4	331.6
2012	333.1	334.6	337.1	338.1	337.7	337.3	336.7	338.6	340.1	340.0	338.4	337.5
2013	338.6	341.4	342.2	341.9	342.5	343.3	343.5	343.9	344.3	343.4	342.8	342.8
2014	344.1	345.4	347.6	348.7	350.0	350.7	350.5	349.9	350.2	349.3	347.5	345.6
2015	344.0	345.5	347.5	348.3	350.0	351.3	351.3	350.8	350.3	350.2	349.4	348.3
2016	348.8	349.1	350.6	352.3	353.7	354.9	354.3	354.7	355.5	356.0	355.4	355.5
2017	357.6	358.7	359.0	360.1	360.4	360.7	360.5	361.6	363.5	363.2	363.3	363.1
2018	365.0	366.7	367.5	369.0	370.5	371.1	371.1	371.3	371.8	372.4	371.2	370.0

Appendix I

Monthly employment level in construction (ELC) from 2003 to 2018

	Jan	Feb	Mar	Apr	May	Jun	Jul	Aug	Sep	Oct	Nov	Dec
2003	5124	5103	5058	5094	5106	5113	5119	5140	5164	5149	5165	5180
2004	5193	5199	5236	5234	5280	5295	5313	5334	5359	5402	5403	5441
2005	5403	5468	5486	5565	5580	5609	5622	5648	5680	5721	5781	5780
2006	5843	5888	5911	5945	5938	5926	5925	5934	5917	5882	5857	5877
2007	5919	5808	5909	5899	5906	5950	5939	5887	5870	5866	5833	5797
2008	5781	5739	5721	5650	5612	5567	5519	5491	5428	5367	5229	5138
2009	5013	4920	4792	4672	4633	4556	4490	4421	4369	4316	4302	4276
2010	4233	4160	4192	4200	4169	4164	4145	4163	4154	4172	4179	4144
2011	4125	4131	4140	4143	4170	4175	4194	4195	4222	4224	4236	4244
2012	4256	4237	4242	4242	4216	4225	4233	4244	4248	4260	4267	4298
2013	4331	4365	4377	4373	4400	4422	4426	4445	4466	4488	4510	4489
2014	4529	4541	4562	4596	4611	4629	4663	4682	4708	4719	4719	4740
2015	4761	4780	4763	4799	4832	4850	4866	4885	4901	4939	4976	5022
2016	5004	5013	5039	5051	5052	5055	5070	5070	5099	5118	5136	5148
2017	5166	5183	5192	5201	5201	5220	5216	5243	5258	5264	5283	5310
2018	5330	5376	5379	5396	5430	5441	5454	5467	5475	5492	5491	5517

Appendix J

Monthly money supply (MS) from 2003 to 2018

	Jan	Feb	Mar	Apr	May	Jun	Jul	Aug	Sep	Oct	Nov	Dec
2003	1227.3	1238.2	1239.3	1250.0	1268.8	1281.0	1287.5	1296.4	1297.2	1297.8	1299.1	1306.2
2004	1306.0	1321.4	1328.7	1332.8	1333.3	1342.7	1340.8	1354.3	1362.5	1362.3	1374.2	1376.0
2005	1367.1	1371.1	1370.8	1358.4	1366.0	1380.1	1369.0	1378.0	1378.6	1376.6	1375.9	1374.3
2006	1379.5	1378.4	1383.1	1381.4	1387.2	1375.5	1372.5	1372.6	1364.3	1370.4	1370.5	1366.4
2007	1371.7	1362.9	1366.4	1378.2	1381.0	1368.2	1371.9	1376.8	1375.6	1379.6	1371.2	1373.0
2008	1377.7	1380.4	1388.6	1391.5	1393.6	1404.6	1421.2	1407.9	1462.2	1474.1	1513.6	1601.1
2009	1582.8	1567.1	1578.8	1612.0	1617.9	1659.0	1662.7	1660.7	1665.1	1678.3	1681.3	1691.9
2010	1674.6	1699.7	1711.9	1699.6	1710.6	1731.7	1724.4	1749.3	1766.5	1780.9	1826.2	1835.8
2011	1853.4	1872.5	1891.3	1901.4	1938.7	1956.2	2001.9	2113.8	2128.1	2140.2	2164.1	2163.5
2012	2201.8	2209.9	2229.0	2248.8	2257.2	2277.3	2318.5	2348.9	2390.1	2422.5	2421.2	2460.6
2013	2473.3	2472.4	2480.4	2515.8	2530.7	2531.4	2545.8	2552.1	2584.5	2623.0	2623.1	2664.4
2014	2696.4	2726.4	2754.8	2778.6	2795.2	2829.5	2841.6	2802.9	2862.2	2868.8	2884.9	2940.7
2015	2941.0	3007.1	2999.6	3000.9	2986.6	3020.0	3039.7	3028.5	3044.2	3018.6	3081.7	3094.9
2016	3099.0	3130.4	3151.1	3200.3	3238.7	3245.7	3247.9	3315.1	3326.5	3335.7	3354.1	3342.4
2017	3390.9	3404.8	3445.5	3454.2	3517.3	3525.9	3550.9	3580.7	3574.2	3606.7	3630.6	3612.0
2018	3653.1	3622.5	3656.3	3660.2	3654.6	3655.1	3676.9	3679.9	3703.7	3720.8	3700.1	3751.3

© The Editor(s) (if applicable) and The Author(s), under exclusive license to Springer Nature Switzerland AG 2023
M. Shahandashti et al., *Construction Analytics*,
https://doi.org/10.1007/978-3-031-27292-9

Appendix K

Monthly US crude oil price (COP) from 2003 to 2018 (dollar/barrel)

	Jan	Feb	Mar	Apr	May	Jun	Jul	Aug	Sep	Oct	Nov	Dec
2003	28.42	31.85	30.1	25.45	24.95	26.84	27.52	27.94	25.23	26.53	27.21	28.53
2004	30.35	31.21	32.86	33.2	35.73	34.53	36.54	40.1	40.56	46.14	42.85	38.22
2005	40.18	42.19	47.56	47.26	44.03	49.83	53.35	58.9	59.64	56.99	53.2	53.24
2006	57.85	55.69	55.64	62.52	64.4	64.65	67.71	67.21	59.37	53.26	52.42	55.03
2007	49.32	52.94	54.95	58.2	58.9	62.35	69.23	67.77	73.27	79.32	87.16	85.28
2008	87.06	89.41	98.44	106.64	118.55	127.47	128.08	112.83	98.5	73.18	53.67	36.8
2009	35	34.14	42.45	45.19	52.67	63.09	60.44	65.28	65.28	69.82	71.99	70.42
2010	72.87	72.74	75.77	78.8	70.91	70.77	71.37	72.07	71.23	76.02	79.2	83.98
2011	85.66	86.69	99.19	108.8	102.46	97.3	97.82	89	90.22	92.28	100.18	98.71
2012	98.99	102.04	105.42	103.62	95.57	83.59	86.1	92.53	95.98	92.24	89.64	89.81
2013	95	95.01	95.54	94.41	94.75	93.82	101.41	102.96	102.32	96.18	88.7	91.85
2014	89.57	96.86	96.17	96.49	95.74	98.68	96.7	90.72	86.87	78.84	71.07	54.86
2015	43.06	44.35	42.66	49.3	54.38	55.88	47.7	39.98	41.6	42.34	38.19	32.26
2016	27.02	25.52	31.87	35.59	41.02	43.96	40.71	40.46	40.55	45	41.65	47.12
2017	48.19	49.41	46.39	47.23	45.19	42.17	43.42	44.96	47.17	49.12	55.19	56.98
2018	62.25	61.18	60.68	63.5	66.16	62.8	67	62.64	63.54	65.18	55.65	47.63

© The Editor(s) (if applicable) and The Author(s), under exclusive license to Springer Nature Switzerland AG 2023
M. Shahandashti et al., *Construction Analytics*,
https://doi.org/10.1007/978-3-031-27292-9

Appendix L

California Construction Cost Index (CCCI) from 2006 to 2018

	Jan	Feb	Mar	Apr	May	Jun	Jul	Aug	Sep	Oct	Nov	Dec
2006	4620	4604	4597	4600	4599	4593	4609	4616	4619	4867	4891	4877
2007	4869	4868	4871	4872	4886	4842	4849	4851	4942	4943	4978	4981
2008	4983	4983	4999	5004	5023	5065	5135	5142	5194	5393	5375	5322
2009	5309	5295	5298	5296	5288	5276	5263	5265	5264	5259	5259	5262
2010	5260	5262	5268	5270	5378	5394	5401	5401	5381	5591	5599	5596
2011	5592	5624	5627	5636	5637	5643	5654	5667	5668	5675	5680	5680
2012	5683	5683	5738	5740	5755	5754	5750	5778	5777	5780	5779	5768
2013	5774	5782	5777	5786	5796	5802	5804	5801	5802	5911	5903	5901
2014	5898	5896	5953	5956	5957	5961	5959	5959	5959	5969	5981	5977
2015	6073	6077	6069	6062	6069	6055	6055	6055	6113	6114	6109	6108
2016	6106	6132	6248	6249	6240	6238	6245	6244	6267	6343	6344	6373
2017	6373	6373	6373	6461	6455	6470	6474	6620	6620	6596	6596	6596
2018	6596	6596	6596	6596	6596	6598	6643	6613	6674	6679	6679	6684

Appendix M

Sim1 time series from 2010 to 2018

	Jan	Feb	Mar	Apr	May	Jun	Jul	Aug	Sep	Oct	Nov	Dec
2010	−1	−2	−1	0	−1	0	1	2	3	2	1	0
2011	1	0	1	0	1	2	1	2	3	2	3	2
2012	1	0	−1	−2	−1	−2	−3	−2	−3	−4	−3	−2
2013	−1	−2	−1	−2	−1	0	1	2	3	4	3	2
2014	3	4	3	4	3	2	1	0	−1	0	1	0
2015	1	0	−1	−2	−1	−2	−3	−2	−3	−2	−3	−2
2016	−3	−4	−5	−4	−3	−4	−3	−2	−3	−2	−3	−4
2017	−3	−4	−3	−4	−5	−6	−7	−8	−7	−6	−5	−4
2018	−5	−6	−5	−4	−3	−4	−5	−4	−3	−4	−5	−6

Appendix N

Sim2 time series from 2010 to 2018

	Jan	Feb	Mar	Apr	May	Jun	Jul	Aug	Sep	Oct	Nov	Dec
2010	1	0	−1	0	−1	−2	−3	−4	−3	−4	−3	−2
2011	−3	−2	−3	−4	−5	−6	−7	−6	−5	−4	−3	−4
2012	−5	−4	−3	−4	−3	−4	−3	−4	−5	−4	−5	−4
2013	−3	−2	−1	0	−1	−2	−3	−2	−3	−4	−5	−6
2014	−7	−6	−7	−6	−7	−8	−9	−8	−9	−10	−11	−10
2015	−11	−12	−13	−14	−13	−12	−11	−12	−13	−14	−15	−14
2016	−13	−12	−13	−12	−13	−12	−13	−14	−13	−14	−15	−14
2017	−15	−16	−15	−16	−17	−18	−19	−18	−17	−16	−17	−16
2018	−17	−16	−15	−16	−17	−18	−19	−20	−19	−18	−17	−18

Appendix O

Complete Results of R Code 7.3:

```
> summary(fn)
Fitting of the distribution ' norm ' by maximum likelihood
Parameters :
       estimate Std. Error
mean 2328.03752  0.8312299
sd     83.12297  0.5877683
Loglikelihood:  -58392.6   AIC:  116789.2   BIC:  116803.6
Correlation matrix:
     mean sd
mean   1  0
sd     0  1
> summary(fu)
Fitting of the distribution ' unif ' by maximum likelihood
Parameters :
    estimate Std. Error
min 2062.559         NA
max 2693.766         NA
Loglikelihood:  -64476.35   AIC:  128956.7   BIC:  128971.1
Correlation matrix:
[1] NA
> summary(ft)
Fitting of the distribution ' triang ' by maximum goodness-of-fit
Parameters :
     estimate
min   2127.398
max   2528.621
mode  2327.775
Loglikelihood:  -Inf    AIC:  Inf    BIC:  Inf
> gofstat(fn,fitnames = "norm")
Goodness-of-fit statistics
                                  norm
Kolmogorov-Smirnov statistic  0.01647556
Cramer-von Mises statistic    0.75125673
Anderson-Darling statistic    4.72742039
```

```
Goodness-of-fit criteria
                                       norm
Akaike's Information Criterion  116789.2
Bayesian Information Criterion  116803.6
> gofstat(fu,fitnames = "unif")
Goodness-of-fit statistics
                                       unif
Kolmogorov-Smirnov statistic      0.31828
Cramer-von Mises statistic      321.65004
Anderson-Darling statistic            Inf

Goodness-of-fit criteria
                                       unif
Akaike's Information Criterion  128956.7
Bayesian Information Criterion  128971.1
> gofstat(ft,fitnames = "triangle")
Goodness-of-fit statistics
                                    triangle
Kolmogorov-Smirnov statistic     0.009774752
Cramer-von Mises statistic       0.244684812
Anderson-Darling statistic               Inf

Goodness-of-fit criteria
                                    triangle
Akaike's Information Criterion           Inf
Bayesian Information Criterion           Inf
```

Appendix O

Complete Results of R Code 7.4:

```
> library(triangle)
> library(fitdistrplus)
> require(mc2d)
> ##General inputs
> set.seed(22)
> n_loop = 1000
> n = 35
> comp1_A = rnorm(n_loop, mean=0.5, sd=0.2)
> comp2_A = rweibull(n_loop, shape=0.5, scale=0.2)
> comp3_A = rtriangle(n_loop, a =80, b = 130, c = 120)
> comp4_A = rep(5, length.out=n_loop)
> recurring_comp= rnorm(n, mean=0.5, sd=0.2)
> Project_A = LCCA_MC(comp1 = comp1_A, comp2 = comp2_A, comp3 = comp3_A, comp4 =
comp4_A, r=0.025, n_loop=n_loop, n=n, recurring_comp = recurring_comp)
    [1] 114.83503  99.21683 138.85488 133.56507 119.23297 135.58122 141.73788
137.38899 142.35397 130.70659
   [11] 109.33912 112.01678 131.54864 135.52752 122.96058 123.51172 136.32969
118.11677 138.64042 111.29508
   [21] 127.24295 143.15267 122.90209 132.43019 136.91494 111.46738 127.55212
121.41822 113.93459 128.54956
   [31] 111.55900 136.18046 131.39493 129.73809 111.61772 110.04892 142.76092
111.95731 122.15015 128.36567
   [41] 133.13735 122.26085 142.82271 122.92310 137.52336 141.88983 136.68785
111.20065 130.15528 135.47655
   [51] 131.32505 133.96651 119.74361 131.23316 124.60056 121.95535 135.92041
135.67040 119.09673 132.05197
   [61] 138.95955 134.92396 126.25243 130.10803 118.06864 108.68122 135.82015
142.90621 135.78790 142.54284
   [71] 141.39187 138.99759 140.39232 113.75713 129.15605 127.25694 125.76272
118.65888 130.90382 125.73048
   [81] 133.69960 144.59145 129.92395 135.05929 143.65252 104.68820 130.19291
139.67764 129.36861 138.70934
   [91]  98.73786 111.67225 138.16314 123.44464 144.97221 133.23189 118.33102
123.79737 137.87075 109.22627
  [101] 126.38073 112.74358 113.84852 134.59159 138.16777 136.17495 135.22751
123.55801 126.04954 129.27508
  [111] 115.99752 124.89699 140.87921 107.04461 132.03394 100.71436 125.07508
129.44671 131.98725 112.32618
  [121] 133.01146 132.34158 108.99901 138.77345 130.23845 139.03710 113.61123
135.73056 110.30103 136.57367
  [131] 111.97624 137.90181 142.80704 114.42367 115.43905 109.87725 115.62284
135.84998 137.58794 122.43133
  [141] 123.53180 127.77128 120.10875 100.99073 112.93212 141.61443 121.39204
134.12377 125.84731 124.92184
  [151] 135.57549 138.00778 136.76059 109.50008 127.05562 126.41575 139.33467
140.64512 136.21143 133.74348
  [161] 116.82144 122.87100 145.60158 105.73688 123.42436 131.22765 109.77733
122.35495 114.39295 131.19552
  [171] 102.80649 128.31099 131.40182 120.40043 140.76267 138.63429 125.32927
121.16865 112.36950 121.40507
  [181] 117.36031 128.96321 135.27897 119.84903 135.51186 131.21182 111.13751
133.62985 136.36602 135.65122
```

[191] 128.68604 134.22415 129.35915 133.99859 135.95963 129.38204 120.54991 118.73377 128.13018 135.30082
[201] 130.21667 127.57078 130.64828 141.28759 135.52233 145.34740 131.23128 115.69555 143.05051 121.66250
[211] 129.28599 123.77521 142.93653 141.18694 144.56498 140.04639 143.56240 104.01257 131.04705 126.56476
[221] 115.76232 123.60228 131.60160 117.75721 145.71317 136.51535 111.86915 135.37543 128.68121 119.66086
[231] 132.80367 126.43776 128.78538 131.20421 125.08228 134.12106 124.29134 134.25426 133.18352 139.49435
[241] 117.99337 109.54761 133.54734 140.60865 132.98030 101.89897 130.55578 129.27387 135.96407 123.87846
[251] 138.56875 138.31370 115.24556 126.59092 122.39216 122.02665 100.74369 130.05670 132.31181 141.55491
[261] 128.25543 137.07730 139.33760 136.01047 137.64325 122.40946 135.96205 127.85067 136.26655 125.69927
[271] 130.46563 131.51536 141.19346 119.27944 107.84339 110.40146 102.86738 122.52655 143.01680 137.57494
[281] 124.37020 138.23842 133.38108 113.59213 139.39232 128.03320 122.39465 136.32377 137.14768 140.19525
[291] 121.79894 140.24168 142.71103 127.03070 115.77253 135.47614 112.04457 139.52886 135.50790 137.49278
[301] 115.48856 119.11902 127.81417 112.75201 140.30919 120.91547 129.58740 107.49976 114.26679 129.22672
[311] 118.19422 131.82290 123.34726 114.86057 140.57586 110.10615 112.44584 140.01607 125.60122 132.54099
[321] 132.11530 134.43383 129.81542 124.19515 131.64308 142.94949 119.51432 137.11211 112.57284 115.03902
[331] 129.00503 122.34939 128.43975 138.17418 121.22662 117.77753 134.82479 126.17608 138.97455 147.17710
[341] 133.35833 142.13881 122.09448 122.02960 131.37085 131.82076 123.39520 135.06876 129.38731 123.31292
[351] 122.62003 133.87172 126.12714 141.36617 115.50845 140.53580 129.07281 136.60065 121.44833 143.07681
[361] 136.87748 139.65852 114.82397 107.25900 112.25151 118.04868 126.86788 125.86633 136.67330 131.51680
[371] 122.64836 139.78102 133.45662 146.27363 120.51632 130.45406 120.77994 137.78271 109.41155 130.05731
[381] 121.14023 133.08288 119.37243 117.02304 121.43781 127.48987 138.38627 118.72343 118.66767 119.72802
[391] 99.18972 116.55332 104.07205 143.27831 124.26159 121.97271 114.30039 142.69351 134.29228 119.68580
[401] 137.38378 109.83497 142.03638 117.20326 106.47683 131.62274 134.52709 134.23993 104.40420 133.18045
[411] 133.81573 133.82324 132.93165 144.82784 134.47209 135.31405 110.62034 139.85119 105.40783 131.76669
[421] 122.07520 132.43507 127.45074 127.13910 128.54822 110.93153 123.27641 132.21927 117.19232 108.71176
[431] 118.26797 129.13845 132.66331 127.12295 137.48945 139.53226 133.79334 125.75613 127.14305 138.07198
[441] 126.06021 141.59536 104.25157 135.48312 140.48800 122.27820 120.77047 134.58471 139.70622 140.94363

Appendix O

[451] 135.65775 142.88503 132.21674 137.53649 125.93705 121.08996 128.91128 126.48857 111.42525 123.52424
[461] 114.83763 137.90701 128.65141 113.25537 126.94139 114.35815 136.41845 142.55073 101.73407 123.84183
[471] 136.89200 132.37237 136.90818 130.17197 131.89949 134.79230 111.64384 139.92504 130.57173 143.78670
[481] 127.52295 142.30373 114.95904 105.81376 133.49442 128.70561 112.97935 137.30428 138.36974 110.52158
[491] 118.18995 129.73692 140.37986 135.81652 126.96146 120.68579 132.66529 136.15437 128.05096 129.25653
[501] 112.61675 129.69762 137.73516 131.59583 124.22129 127.83415 137.29757 126.11545 104.50387 109.52269
[511] 97.79631 127.27086 117.20082 133.04810 133.50761 127.32229 140.45212 138.50709 126.48482 135.35141
[521] 134.60288 131.99251 115.96442 117.65126 123.83602 122.03359 136.91322 133.00864 126.10276 128.67085
[531] 138.02369 122.49131 141.16766 116.06420 137.19931 142.47566 136.54706 141.53452 134.07899 126.01952
[541] 138.00313 127.86824 114.80870 140.44914 141.99463 131.92720 116.06942 109.25840 110.25912 107.81079
[551] 121.43649 108.79639 134.42983 111.88990 114.07463 135.72987 129.84268 128.31440 137.32630 139.18724
[561] 127.95849 117.20865 139.38685 110.90128 130.75051 133.63449 119.80081 125.50551 115.55065 137.33044
[571] 109.16242 131.29669 132.63186 132.80127 134.10067 140.02912 137.35503 138.53562 136.98635 133.18394
[581] 139.30754 140.47583 134.32467 122.67783 134.80403 126.83906 133.85607 136.59115 121.35869 127.62519
[591] 125.54969 129.73802 132.84519 138.45548 134.98836 139.03761 136.12884 130.96282 143.71824 138.12849
[601] 135.64423 109.23705 130.49629 122.44331 119.13321 134.16956 104.33571 134.19005 136.09722 137.81322
[611] 141.58870 128.78262 112.34211 116.41646 140.45063 135.55047 146.10418 129.75792 126.91668 102.30537
[621] 136.11544 131.56748 131.91693 110.71876 127.67734 133.27934 105.80703 131.76712 120.37131 113.44488
[631] 136.10465 136.67536 126.29020 105.71676 138.96323 124.23770 130.09864 121.17152 131.61676 128.98583
[641] 105.26421 136.68605 127.68438 132.62653 134.05050 115.93372 124.24622 122.13384 127.25565 104.00072
[651] 131.95766 118.38572 144.65479 119.40502 119.87395 122.34865 137.87523 128.73227 131.44286 119.69298
[661] 135.49498 128.76349 118.57693 126.85052 108.59198 135.99591 124.07597 134.66910 124.77455 126.56310
[671] 138.67267 136.20862 131.41188 127.81634 125.52144 133.29409 130.55186 136.95974 143.54128 143.40380
[681] 134.99863 135.03989 128.30629 132.05170 111.14562 144.28704 138.10256 136.63426 124.70003 139.73873
[691] 124.15859 115.19786 132.94694 133.93818 99.65595 140.85575 119.24761 121.15932 141.82886 119.57789
[701] 107.11374 144.76302 110.23656 112.06066 133.46018 128.26584 119.35849 106.49246 137.73368 131.97573

[711] 133.55517 141.40885 130.23216 122.52559 137.79803 125.02028 118.28995 116.85126 126.88322 143.10191
[721] 123.17931 138.87523 134.33352 115.69178 141.53919 133.76212 134.55498 130.45512 129.11962 122.44000
[731] 125.42189 121.32301 121.74165 144.16005 138.36141 132.17442 132.56446 137.76009 119.36862 123.79636
[741] 111.17125 137.06485 119.14029 114.38166 110.91944 124.35562 128.25295 140.15194 141.58730 104.22203
[751] 134.64655 123.91637 131.59720 141.43797 127.14720 128.87181 131.38407 144.47779 135.77559 124.14442
[761] 126.93625 126.66905 103.59900 135.02921 138.07374 136.59844 132.76148 111.21241 122.97261 127.05954
[771] 124.85436 108.83670 121.47822 125.45965 125.71332 115.62876 115.06274 137.32693 108.31863 126.13912
[781] 112.61072 137.58583 118.27818 134.01795 110.72055 143.81128 121.47179 101.81378 134.18948 131.91171
[791] 124.43510 133.91939 103.89965 107.53413 118.55654 135.69137 120.26163 111.08235 139.81453 142.62193
[801] 132.62055 136.60056 117.63936 116.09538 132.46714 134.47695 127.12869 127.64945 144.17239 123.47327
[811] 121.35583 123.99204 104.34286 122.30374 135.29461 123.82607 139.48110 128.88808 136.13167 135.09171
[821] 122.98771 125.48803 131.86480 111.02346 127.57571 104.54410 111.76789 119.93353 128.51330 144.11497
[831] 125.29149 145.16032 110.87660 130.29175 128.93802 125.92467 141.06856 138.00894 116.18929 121.53350
[841] 117.79096 129.79387 125.87946 124.97491 108.35095 139.24982 122.42487 130.38288 131.26329 112.81601
[851] 136.27888 132.99137 116.20919 121.34209 141.14424 135.04393 141.49466 105.37505 115.55955 126.13390
[861] 136.54413 118.39964 121.64966 110.80154 113.82671 132.91691 132.26058 123.53415 103.95419 142.67551
[871] 123.76547 125.23665 128.14552 123.07420 115.53020 137.48639 117.36042 141.18906 125.66574 98.91886
[881] 142.87184 129.63408 134.03504 134.96522 112.56922 135.28749 138.05586 146.76281 126.62651 121.21038
[891] 112.81448 130.77237 130.41193 126.54131 134.12284 135.73346 137.21318 127.35147 126.77032 111.07945
[901] 135.41568 135.94035 143.78257 112.63829 112.34298 117.33709 105.67535 128.18258 134.07045 109.99963
[911] 135.51283 120.72756 131.25494 122.09551 125.32617 118.35735 139.42022 101.44104 122.46277 120.59761
[921] 134.95032 118.11847 136.68680 134.74350 134.10786 127.19645 127.56028 132.05176 132.60956 135.88678
[931] 107.00318 131.24387 114.97427 101.87118 128.24251 136.61853 114.57940 129.25498 124.88896 117.72914
[941] 139.53247 138.13800 134.67019 127.58218 119.64044 139.55661 114.46061 123.19595 112.99441 124.63456
[951] 122.57644 128.82975 132.41072 134.58769 140.89941 99.71670 146.16637 106.63741 136.20047 112.43065
[961] 123.28488 112.08104 130.04627 119.74201 138.06372 130.89900 121.11455 131.46764 133.31084 133.04728

```
 [971] 144.12880 119.56205 122.11261 113.33856 120.20314 108.10879 132.92117
 143.92220 123.01320 129.80810
 [981] 131.14509 122.54087 135.33557 137.85726 124.88813  99.26073 136.04181
 131.08085 120.94603 113.89586
 [991]  97.50048 133.54245 126.78576 130.89058 111.60630 102.98953 104.61844
 134.37796 126.24674 139.06918
> # Fit distribution to get the parameters.
> fn <- fitdist(Project_A, "norm")
$start.arg
$start.arg$mean
[1] 127.2699

$start.arg$sd
[1] 10.79468


$fix.arg
NULL

> summary(fn)
Fitting of the distribution ' norm ' by maximum likelihood
Parameters :
      estimate Std. Error
mean 127.26991  0.3413578
sd    10.79468  0.2413764
Loglikelihood:  -3797.992   AIC:  7599.984   BIC:  7609.8
Correlation matrix:
     mean sd
mean    1  0
sd      0  1

> fu <- fitdist(Project_A, "unif")
$start.arg
$start.arg$min
[1] 0

$start.arg$max
[1] 1


$fix.arg
NULL

> summary(fu)
Fitting of the distribution ' unif ' by maximum likelihood
Parameters :
    estimate Std. Error
min  97.50048         NA
max 147.17710         NA
Loglikelihood:  -3905.534   AIC:  7815.069   BIC:  7824.884
Correlation matrix:
[1] NA
```

```
> ft <- fitdist(Project_A, "triang", method="mge", start = list(min=93, max=153,
mode=135), gof ='CvM')
$start.arg
$start.arg$min
[1] 93

$start.arg$max
[1] 153

$start.arg$mode
[1] 135


$fix.arg
NULL

> summary(ft)
Fitting of the distribution ' triang ' by maximum goodness-of-fit
Parameters :
       estimate
min     98.10512
max    147.10981
mode   136.87340
Loglikelihood:  -Inf   AIC:  Inf   BIC:  Inf
> par(mfrow = c(2, 2))
> plot.legend <- c("normal","uniform", "triangular")
> denscomp(list(fn,fu, ft), legendtext = plot.legend)
> qqcomp(list(fn,fu, ft), legendtext = plot.legend)
> cdfcomp(list(fn,fu, ft), legendtext = plot.legend)
> ppcomp(list(fn,fu, ft), legendtext = plot.legend)
> # Conduct the Fit Test to see which one is the best fit.
> gofstat(fn,fitnames = "norm")
Goodness-of-fit statistics
                                  norm
Kolmogorov-Smirnov statistic   0.0799443
Cramer-von Mises statistic     1.6857369
Anderson-Darling statistic    10.3898521

Goodness-of-fit criteria
                                    norm
Akaike's Information Criterion   7599.984
Bayesian Information Criterion   7609.800
> gofstat(fu,fitnames = "unif")
Goodness-of-fit statistics
                                   unif
Kolmogorov-Smirnov statistic    0.2118609
Cramer-von Mises statistic     17.5869812
Anderson-Darling statistic            Inf

Goodness-of-fit criteria
                                    unif
Akaike's Information Criterion   7815.069
Bayesian Information Criterion   7824.884
```

Appendix O

```
> gofstat(ft,fitnames = "triangle")
Goodness-of-fit statistics
                                triangle
Kolmogorov-Smirnov statistic  0.01930999
Cramer-von Mises statistic    0.04195269
Anderson-Darling statistic           Inf

Goodness-of-fit criteria
                                triangle
Akaike's Information Criterion       Inf
Bayesian Information Criterion       Inf
```

Complete Results of R Code 7.6:

```
> descdist(Project_A)
summary statistics
------
min:   -0.507964    max:   1.933197
median:   0.8592928
mean:   0.8081595
estimated sd:   0.5435923
estimated skewness:   -0.3450083
estimated kurtosis:   2.420585

> descdist(Project_B)
summary statistics
------
min:   -0.01144907   max:   1.376857
median:   0.8841355
mean:   0.8237969
estimated sd:   0.3221833
estimated skewness:   -0.5143923
estimated kurtosis:   2.334869

> fn_A <- fitdist(Project_A, "norm")
$start.arg
$start.arg$mean
[1] 0.8081595
$start.arg$sd
[1] 0.5433204
$fix.arg
NULL

> summary(fn_A)
Fitting of the distribution ' norm ' by maximum likelihood
Parameters :
      estimate Std. Error
mean 0.8081595 0.01718130
sd   0.5433204 0.01214883
Loglikelihood:  -808.8825   AIC:  1621.765   BIC:  1631.58
Correlation matrix:
     mean sd
mean   1  0
sd     0  1


> fu_A <- fitdist(Project_A, "unif")
$start.arg
$start.arg$min
[1] 0
$start.arg$max
[1] 1
$fix.arg
NULL

> summary(fu_A)
Fitting of the distribution ' unif ' by maximum likelihood
Parameters :
     estimate Std. Error
min -0.507964        NA
max  1.933197        NA
Loglikelihood:  -892.4739   AIC:  1788.948   BIC:  1798.763
Correlation matrix:
[1] NA
```

Appendix O

```
> ft_A <- fitdist(Project_A, "triang", method="mge", start = list(min=-0.5, max=1.5,
mode=0.6), gof = "CvM")
$start.arg
$start.arg$min
[1] -0.5
$start.arg$max
[1] 1.5
$start.arg$mode
[1] 0.6
$fix.arg
NULL

> summary(ft_A)
Fitting of the distribution ' triang ' by maximum goodness-of-fit
Parameters :
       estimate
min  -0.6542849
max   1.9283880
mode  1.1492213
Loglikelihood:  -Inf   AIC:  Inf   BIC:  Inf

> fn_B <- fitdist(Project_B, "norm")
$start.arg
$start.arg$mean
[1] 0.8237969
$start.arg$sd
[1] 0.3220221
$fix.arg
NULL

> summary(fn_B)
Fitting of the distribution ' norm ' by maximum likelihood
Parameters :
      estimate  Std. Error
mean 0.8237969 0.010183234
sd   0.3220221 0.007200321
Loglikelihood:  -285.8035   AIC:  575.6071   BIC:  585.4226
Correlation matrix:
             mean           sd
mean 1.000000e+00 1.041976e-12
sd   1.041976e-12 1.000000e+00

> fu_B <- fitdist(Project_B, "unif")
$start.arg
$start.arg$min
[1] 0
$start.arg$max
[1] 1
$fix.arg
NULL

> summary(fu_B)
Fitting of the distribution ' unif ' by maximum likelihood

Parameters :
       estimate Std. Error
min -0.01144907         NA
max  1.37685655         NA
Loglikelihood:  -328.084   AIC:  660.168   BIC:  669.9836
Correlation matrix:
[1] NA
```

```
> ft_B <- fitdist(Project_B, "triang", method="mge", start = list(min=-0.5, max=1.5,
mode=0.6) , gof = "CvM")
$start.arg
$start.arg$min
[1] -0.5
$start.arg$max
[1] 1.5
$start.arg$mode
[1] 0.6
$fix.arg
NULL

> summary(ft_B)
Fitting of the distribution ' triang ' by maximum goodness-of-fit
Parameters :
        estimate
min   -0.07958851
max    1.36952972
mode   1.18104993
Loglikelihood:  -Inf   AIC:  Inf   BIC:  Inf

> gofstat(fn_A,fitnames = "norm")
Goodness-of-fit statistics                              norm
Kolmogorov-Smirnov statistic 0.05574385
Cramer-von Mises statistic   0.78665868
Anderson-Darling statistic   5.22607547
Goodness-of-fit criteria
                                         norm
Akaike's Information Criterion 1621.765
Bayesian Information Criterion 1631.580

> gofstat(fu_A,fitnames = "unif")
Goodness-of-fit statistics
                                  unif
Kolmogorov-Smirnov statistic 0.148105
Cramer-von Mises statistic   7.723826
Anderson-Darling statistic        Inf
Goodness-of-fit criteria
                                  unif
Akaike's Information Criterion 1788.948
Bayesian Information Criterion 1798.763

> gofstat(ft_A,fitnames = "triangle")
Goodness-of-fit statistics
                                triangle
Kolmogorov-Smirnov statistic 0.02200626
Cramer-von Mises statistic   0.05253836
Anderson-Darling statistic        Inf
Goodness-of-fit criteria
                               triangle
Akaike's Information Criterion      Inf
Bayesian Information Criterion      Inf
```

Appendix O

```
> gofstat(fn_B,fitnames = "norm")
Goodness-of-fit statistics
                                       norm
Kolmogorov-Smirnov statistic     0.08688968
Cramer-von Mises statistic       2.11427307
Anderson-Darling statistic      13.28722309
Goodness-of-fit criteria
                                      norm
Akaike's Information Criterion    575.6071
Bayesian Information Criterion    585.4226

> gofstat(fu_B,fitnames = "unif")
Goodness-of-fit statistics
                                     unif
Kolmogorov-Smirnov statistic     0.193234
Cramer-von Mises statistic      15.945081
Anderson-Darling statistic            Inf
Goodness-of-fit criteria
                                     unif
Akaike's Information Criterion   660.1680
Bayesian Information Criterion   669.9836

> gofstat(ft_B,fitnames = "triangle")
Goodness-of-fit statistics
                                   triangle
Kolmogorov-Smirnov statistic     0.01685045
Cramer-von Mises statistic       0.04643023
Anderson-Darling statistic              Inf
Goodness-of-fit criteria
                                   triangle
Akaike's Information Criterion          Inf
Bayesian Information Criterion          Inf
```

Index

A
ARCH effects, 48, 53, 60, 61
ARIMA-ARCH model, 53–58
ARIMA-GARCH model, 56, 57, 59–61
Autocorrelation, 11, 21, 26, 30, 32, 33, 35–39, 41, 46, 49, 53–54, 63
Autoregressive, 16, 18–22, 25, 28, 29, 37–38, 41, 46–53, 69
Autoregressive conditional heteroscedasticity (ARCH) model, 46, 47, 50, 51
Autoregressive integrated moving average (ARIMA), 16, 27–35, 41, 47–52, 56, 57, 59–61, 93
Autoregressive moving average (ARMA), 16, 25–28, 52, 61, 75

B
Binomial decision tree, 111–122

C
Co-integration test, 64, 68, 70, 73
Conditional variance, 52, 61
Construction analytics, 2–5, 8, 64, 110
Construction cost, 2, 3, 8–11, 93, 106, 110, 114, 115, 122, 125, 134, 137, 142, 143, 149, 153, 165
Construction cost forecasting, 2
Construction forecasting, 73
Construction investment valuation under uncertainty, 110

E
Explanatory time series, 64–70
Exponential smoothing (ES), 15, 16, 22–25, 41, 42

F
Forecasting, 2–5, 8, 11, 13, 15–17, 22, 23, 30, 35, 36, 41, 53, 56–61, 63, 64, 66, 69–73, 76–79, 85, 91–93

G
Gated recurrent unit (GRU), 77, 89–93
General ARCH (GARCH) models, 56, 59, 60
Granger causality, 64, 66–67, 70, 73

H
Heteroscedasticity, 35, 37, 39, 41, 42, 46–54, 61
Homoscedasticity, 37–39, 53–55

I
Investment decision-making under uncertainty, 110
Investment valuation, 2–5, 95–144

L
LM-ARCH test, 54
Long short-term memory (LSTM), 77, 84–89, 91–93

M
Machine learning methods, 3
Monte Carlo simulation, 96, 97, 99, 117, 118, 122
Moving average, 8, 16, 18, 25, 28, 41, 52, 91
Multivariate time series models, 63, 64, 73

R
Real options analysis, 3, 110–111, 119, 122
Recurrent neural networks (RNNs), 76–89, 91–93
R Programming, 2

S
Seasonal autoregressive moving average (SARIMA), 29–34, 41, 69–71, 89, 91
Seasonality, 8, 11, 14–16, 22, 23, 29, 30, 41, 63, 84, 88, 91

Stationarity, 8, 11–14, 16, 25–27, 35, 41, 63, 65–66
Stochastic life-cycle cost analysis, 95, 96, 101, 122
Strategic decision-making, 2, 3, 5

T
Time-series volatility model, 46, 49, 53–55, 59
Time-varying variance, 46, 47, 49, 52, 53, 56, 59–61

U
Univariate time series forecasting, 3, 8, 16–35, 41

V
Vector autoregressive (VAR), 63, 69, 70, 73, 93
Vector error correction (VEC), 63, 69–73, 91–93

W
Weighted Ljung-Box tests, 53

SPRINGER NATURE

GPSR Compliance

The European Union's (EU) General Product Safety Regulation (GPSR) is a set of rules that requires consumer products to be safe and our obligations to ensure this.

If you have any concerns about our products, you can contact us on ProductSafety@springernature.com

In case Publisher is established outside the EU, the EU authorized representative is:

Springer Nature Customer Service Center GmbH
Europaplatz 3
69115 Heidelberg, Germany

The manufacturer's authorised representative in the EU is Springer Nature Customer Service Centre GmbH, Europaplatz 3, 69115 Heidelberg, Germany. If you have any concerns regarding our products, please contact ProductSafety@springernature.com

Printed and bound by CPI Group (UK) Ltd, Croydon, CR0 4YY

25/03/2026

02078177-0017